S0-AXK-619

GETTING STARTED WITH STATA
FOR MAC®
RELEASE 12

A Stata Press Publication
StataCorp LP
College Station, Texas

® Copyright © 1985–2011 StataCorp LP
All rights reserved
Version 12

Published by Stata Press, 4905 Lakeway Drive, College Station, Texas 77845
Typeset in TEX
Printed in the United States of America

10 9 8 7 6 5 4 3 2 1

ISBN-10: 1-59718-082-3
ISBN-13: 978-1-59718-082-5

This manual is protected by copyright. All rights are reserved. No part of this manual may be reproduced, stored in a retrieval system, or transcribed, in any form or by any means—electronic, mechanical, photocopy, recording, or otherwise—without the prior written permission of StataCorp LP unless permitted subject to the terms and conditions of a license granted to you by StataCorp LP to use the software and documentation. No license, express or implied, by estoppel or otherwise, to any intellectual property rights is granted by this document.

StataCorp provides this manual "as is" without warranty of any kind, either expressed or implied, including, but not limited to, the implied warranties of merchantability and fitness for a particular purpose. StataCorp may make improvements and/or changes in the product(s) and the program(s) described in this manual at any time and without notice.

The software described in this manual is furnished under a license agreement or nondisclosure agreement. The software may be copied only in accordance with the terms of the agreement. It is against the law to copy the software onto DVD, CD, disk, diskette, tape, or any other medium for any purpose other than backup or archival purposes.

The automobile dataset appearing on the accompanying media is Copyright © 1979 by Consumers Union of U.S., Inc., Yonkers, NY 10703-1057 and is reproduced by permission from CONSUMER REPORTS, April 1979.

Icons other than the Stata icon are licensed from the Iconfactory, Inc. They remain property of the Iconfactory, Inc., and may not be reproduced or redistributed.

Stata, **STaTa** , Stata Press, Mata, **maTa** , and NetCourse are registered trademarks of StataCorp LP.

Stata and Stata Press are registered trademarks with the World Intellectual Property Organization of the United Nations.

NetCourseNow is a trademark of StataCorp LP.

Other brand and product names are registered trademarks or trademarks of their respective companies.

For copyright information about the software, type `help copyright` within Stata.

The suggested citation for this software is

StataCorp. 2011. *Stata: Release 12*. Statistical Software. College Station, TX: StataCorp LP.

Table of Contents

Cross-referencing the documentation

When reading this manual, you will find references to other Stata manuals. For example,

[U] **26 Overview of Stata estimation commands**

[R] **regress**

[D] **reshape**

The first example is a reference to chapter 26, *Overview of Stata estimation commands*, in the *User's Guide*; the second is a reference to the regress entry in the *Base Reference Manual*; and the third is a reference to the reshape entry in the *Data-Management Reference Manual*.

All the manuals in the Stata Documentation have a shorthand notation:

[GSM]	*Getting Started with Stata for Mac*
[GSU]	*Getting Started with Stata for Unix*
[GSW]	*Getting Started with Stata for Windows*
[U]	*Stata User's Guide*
[R]	*Stata Base Reference Manual*
[D]	*Stata Data-Management Reference Manual*
[G]	*Stata Graphics Reference Manual*
[XT]	*Stata Longitudinal-Data/Panel-Data Reference Manual*
[MI]	*Stata Multiple-Imputation Reference Manual*
[MV]	*Stata Multivariate Statistics Reference Manual*
[P]	*Stata Programming Reference Manual*
[SEM]	*Stata Structural Equation Modeling Reference Manual*
[SVY]	*Stata Survey Data Reference Manual*
[ST]	*Stata Survival Analysis and Epidemiological Tables Reference Manual*
[TS]	*Stata Time-Series Reference Manual*
[I]	*Stata Quick Reference and Index*
[M]	*Mata Reference Manual*

Detailed information about each of these manuals may be found online at

http://www.stata-press.com/manuals/

About this manual

This manual discusses **Stata for Mac**®. Stata for Windows® users should see *Getting Started with Stata for Windows*; Stata for Unix® users should see *Getting Started with Stata for Unix*. This manual is intended both for people who are completely new to Stata and for experienced Stata users new to Stata for Mac. Previous Stata users will also find it helpful as a tutorial on some new features in Stata for Mac.

Following the numbered chapters are four appendices with information specific to Stata for Mac.

We provide several types of technical support to registered Stata users. [GSM] **4 Getting help** describes the resources available to help you learn about Stata's commands and features. One of these resources is the Stata website (http://www.stata.com), where you will find answers to frequently asked questions (FAQs), as well as other useful information. If you still have questions after looking at the Stata website and the other resources described in [GSM] **19 Updating and extending Stata—Internet functionality**, you can contact us as described in [U] **3.9 Technical support**.

Using this manual

The new user will get the most out of this book by treating it as an exercise book, working through each example at the computer. The material builds, so material from earlier chapters will often be used in later chapters. Bear in mind that Stata is a rich and deep statistical package—just as statistics itself is rich and deep. The time spent working the examples will be repaid with dividends when doing true statistical analyses.

The experienced user may still have something to learn from this manual, despite its name. We suggest looking through the chapters to see if there is anything new or forgotten.

Notational conventions

The preferences for Stata are located under the Application menu, as is common for applications in Mac OS X. The name of this menu, however, depends on what flavor of Stata you have installed: it could be **Stata/MP 12.0**, **Stata/SE 12.0**, **Stata/IC 12.0**, or **smStata 12.0**. Throughout this documentation, we will call it **Stata**. Thus, we will say that Preferences can be found at **Stata > Preferences**, though the exact name of the top-level menu may be slightly different.

You will see many references to right-clicking in this manual. If you have a one-button mouse, the equivalent action is a *Control*+click.

1   Introducing Stata—sample session

Introducing Stata

This chapter will run through a sample work session, introducing you to a few of the basic tasks that can be done in Stata, such as opening a dataset, investigating the contents of the dataset, using some descriptive statistics, making some graphs, and doing a simple regression analysis. As you would expect, we will only brush the surface of many of these topics. This approach should give you a sample of what Stata can do and how Stata works. There will be brief explanations along the way, with references to chapters later in this book, as well as to the online help and other Stata manuals. We will run through the session by using both menus and dialogs and Stata's commands, so that you can gain some familiarity with them both.

Take a seat at your computer, put on some good music, and work along with the book.

Sample session

The dataset that we will use for this session is a set of data about vintage 1978 automobiles sold in the United States.

To follow along using point-and-click, note that the menu items are given by **Menu > Menu Item > Submenu Item > etc**. To follow along using the Command window, type the commands that follow a dot (.) in the boxed listings below into the small window labeled **Command**. When there is something of note about the structure of a command, it will be pointed out as a "Syntax note".

Start by loading the auto dataset, which is included with Stata. To use the menus,

1. Select **File > Example Datasets...**.
2. Click on Example datasets installed with Stata.
3. Click on use for auto.dta.

The result of this command is threefold:

- The following output appears in the large Results window:

```
. sysuse auto.dta
(1978 Automobile Data)
```

The output consists of a command and its result. The command, sysuse auto.dta, is bold and follows the period (.). The result, (1978 Automobile Data), is in the standard face here and is a brief description of the dataset.

Note: If a command intrigues you, you can type help *commandname* in the Command window to find help. If you want to explore at any time, **Help > Search...** can be informative.

- The same command, sysuse auto.dta, appears in the small Review window to the left. The Review window keeps track of commands Stata has run, successful and unsuccessful. The commands can then easily be rerun. See [GSM] **2 The Stata user interface** for more information.
- A series of variables appears in the small Variables window to the upper right.
- Some information about make, the first variable in the dataset, appears in the small Properties window to the lower right.

1

You could have opened the dataset by typing `sysuse auto` in the Command window and pressing *Return*. Try this now. `sysuse` is a command that loads (uses) example (system) datasets. As you will see during this session, Stata commands are often simple enough that it is faster to use them directly. This will be especially true once you become familiar with the commands you use the most in your daily use of Stata.

Syntax note: In the above example, `sysuse` is the Stata command, whereas `auto` is the name of a Stata data file.

Simple data management

We can get a quick glimpse at the data by browsing it in the **Data Editor**. This can be done by clicking on the **Data Editor (Browse)** button, , or by selecting **Data > Data Editor > Data Editor (Browse)** from the menus, or by typing the command `browse`.

Syntax note: Here the command is `browse` and there are no other arguments.

When the Data Editor opens, you can see that Stata regards the data as one rectangular table. This is true for all Stata datasets. The columns represent *variables*, whereas the rows represent *observations*. The variables have somewhat descriptive names, whereas the observations are numbered.

The data are displayed in multiple colors—at first glance it appears that the variables listed in black are numeric, whereas those that are in colors are text. This is worth investigating. Click on a cell under the `make` variable: the input box at the top displays the make of the car. Scroll to the right until you see the `foreign` variable. Click on one of its cells. Although the cell may display "Domestic", the input box displays a 0. This shows that Stata can store categorical data as numbers but display human-readable text. This is done by what Stata calls *value labels*. Finally, under the `rep78` variable, which looks to be numeric, there are some cells containing just a period (.). The periods correspond to missing values.

Looking at the data in this fashion, though comfortable, lends little information about the dataset. It would be useful for us to get more details about what the data are and how the data are stored. Close the Data Editor by clicking on its close button.

We can see the structure of the dataset by *describing* its contents. This can be done either by going to **Data > Describe data > Describe data in memory** in the menus and clicking on **OK** or by typing describe in the Command window and pressing *Return*. Regardless of which method you choose, you will get the same result:

```
. describe

Contains data from /Applications/Stata/ado/base/a/auto.dta
  obs:            74                          1978 Automobile Data
  vars:           12                          13 Apr 2011 17:45
  size:        3,182                          (_dta has notes)

              storage   display    value
variable name   type    format     label      variable label

make           str18    %-18s                 Make and Model
price          int      %8.0gc                Price
mpg            int      %8.0g                 Mileage (mpg)
rep78          int      %8.0g                 Repair Record 1978
headroom       float    %6.1f                 Headroom (in.)
trunk          int      %8.0g                 Trunk space (cu. ft.)
weight         int      %8.0gc                Weight (lbs.)
length         int      %8.0g                 Length (in.)
turn           int      %8.0g                 Turn Circle (ft.)
displacement   int      %8.0g                 Displacement (cu. in.)
gear_ratio     float    %6.2f                 Gear Ratio
foreign        byte     %8.0g      origin     Car type

Sorted by:  foreign
```

If your listing stops short, and you see a blue —more— at the base of the Results window, pressing the Spacebar or clicking on the blue —more— itself will allow the command to be completed.

At the top of the listing, some information is given about the dataset, such as where it is stored on disk, how much memory it occupies, and when the data were last saved. The bold 1978 Automobile Data is the short description that appeared when the dataset was opened and is referred to as a *data label* by Stata. The phrase _dta has notes informs us that there are notes attached to the dataset. We can see what notes there are by typing notes in the Command window:

```
. notes

_dta:
  1.  from Consumer Reports with permission
```

Here we see a short note about the source of the data.

Looking back at the listing from describe, we can see that Stata keeps track of more than just the raw data. Each variable has the following:

- A *variable name*, which is what you call the variable when communicating with Stata. Variable names are one type of Stata name. See [U] **11.3 Naming conventions**.
- A *storage type*, which is the way in which Stata stores its data. For our purposes, it is enough to know that types beginning with str are *string*, or text, variables, whereas all others are numeric. See [U] **12 Data**.

- A *display format*, which controls how Stata displays the data in tables. See [U] **12.5 Formats: Controlling how data are displayed**.

- A *value label* (possibly). This is the mechanism that allows Stata to store numerical data while displaying text. See [GSM] **9 Labeling data** and [U] **12.6.3 Value labels**.

- A *variable label*, which is what you call the variable when communicating with other people. Stata uses the variable label when making tables, as we will see.

A dataset is far more than simply the data it contains. It is also information that makes the data usable by someone other than the original creator.

Although describing the data tells us something about the structure of the data, it says little about the data themselves. The data can be summarized by clicking on **Statistics > Summaries, tables, and tests > Summary and descriptive statistics > Summary statistics** and clicking on the **OK** button. You could also type summarize in the Command window and press *Return*. The result is a table containing summary statistics about all the variables in the dataset:

```
. summarize
    Variable |       Obs        Mean    Std. Dev.       Min        Max

        make |         0
       price |        74    6165.257    2949.496       3291      15906
         mpg |        74     21.2973    5.785503         12         41
       rep78 |        69    3.405797    .9899323          1          5
    headroom |        74    2.993243    .8459948        1.5          5

       trunk |        74    13.75676    4.277404          5         23
      weight |        74    3019.459    777.1936       1760       4840
      length |        74    187.9324    22.26634        142        233
        turn |        74    39.64865    4.399354         31         51
displacement |        74    197.2973    91.83722         79        425

   gear_ratio |        74    3.014865    .4562871       2.19       3.89
      foreign |        74    .2972973    .4601885          0          1
```

From this simple summary, we can learn a bit about the data. First of all, the prices are nothing like today's car prices—of course, these cars are now antiques. We can see that the gas mileages are not particularly good. Automobile aficionados can gain some feel for other characteristics.

There are two other important items here:

- The make variable is listed as having no observations. It really has no numerical observations because it is a string (text) variable.

- The rep78 variable has five fewer observations than the other numerical variables. This implies that rep78 has five missing values.

Although we could use the summarize and describe commands to get a bird's eye view of the dataset, Stata has a command that gives a good in-depth description of the structure, contents, and values of the variables: the codebook command. Either type codebook in the Command window and press *Return* or navigate the menus to **Data > Describe data > Describe data contents (codebook)** and click on **OK**. We get a large amount of output that is worth investigating. In fact, we get more output than can fit on one screen, as can be seen by the blue —more— at the bottom of the Results window. Press the Spacebar a few times to get all the output to scroll past. (For more about —more—, see *More* in [GSM] **10 Listing data and basic command syntax**.) Look over the output to see that much can be learned from this simple command. You can scroll back in the Results window to see earlier results, if need be. We will focus on the output for make, rep78, and foreign.

To start our investigation, we would like to run the `codebook` command on just one variable, say, `make`. We can do this with menus or the command line, as usual. To get the `codebook` output for `make` with the menus, start by navigating as before, to **Data > Describe data > Describe data contents (codebook)**. When the dialog appears, there are multiple ways to tell Stata to consider only the `make` variable:

- We could type `make` into the *Variables* field.
- The *Variables* field is a combobox control that accepts variable names. Clicking on the drop triangle to the right of the *Variables* field displays a list of the variables from the current dataset. Selecting a variable from the list will, in this case, enter the variable name into the edit field.

A much easier solution is to type `codebook make` in the Command window and then press *Return*. The result is informative:

```
. codebook make

make                                                              Make and Model

                type:  string (str18), but longest is str17
        unique values:  74                       missing "":  0/74
             examples:  "Cad. Deville"
                        "Dodge Magnum"
                        "Merc. XR-7"
                        "Pont. Catalina"

              warning:  variable has embedded blanks
```

The first line of the output tells us the variable name (`make`) and the variable label (`Make and Model`). The variable is stored as a string (which is another way of saying "text") with a maximum length of 18 characters, though a size of only 17 characters would be enough. All the values are unique, so if need be, `make` could be used as an identifier for the observations—something that is often useful when putting together data from multiple sources or when trying to weed out errors from the dataset. There are no missing values, but there are blanks within the makes. This latter fact could be useful if we were expecting `make` to be a one-word string variable.

Syntax note: Telling the `codebook` command to run on the `make` variable is an example of using a *varlist* in Stata's syntax.

Looking at the `foreign` variable can teach us about value labels. We would like to look at the codebook output for this variable, and on the basis of our latest experience, it would be easy to type `codebook foreign` into the Command window (from here on, we will not explicitly say to press the *Return* key) to get the following output:

```
. codebook foreign

foreign                                                                Car type

                type:  numeric (byte)
               label:  origin

               range:  [0,1]                              units:  1
        unique values:  2                             missing .:  0/74

          tabulation:  Freq.   Numeric   Label
                          52         0   Domestic
                          22         1   Foreign
```

We can glean that `foreign` is an indicator variable because its only values are 0 and 1. The variable has a value label that displays `Domestic` instead of 0 and `Foreign` instead of 1. There are two advantages of storing the data in this form:

- Storing the variable as a byte takes less memory because each observation uses 1 byte instead of the 8 bytes needed to store "Domestic". This is important in large datasets. See [U] **12.2.2 Numeric storage types**.

- As an indicator variable, it is easy to incorporate into statistical models. See [U] **25 Working with categorical data and factor variables**.

Finally, we can learn a little about a poorly labeled variable with missing values by looking at the `rep78` variable. Typing `codebook rep78` into the Command window yields

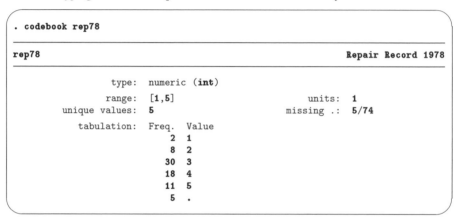

`rep78` appears to be a categorical variable, but because of lack of documentation, we do not know what the numbers mean. (To see how we would label the values, see *Changing data* in [GSM] **6 Using the Data Editor** and see [GSM] **9 Labeling data**.) This variable has five missing values, meaning that there are five observations for which the repair record is not recorded. We could use the Data Editor to investigate these five observations, but we will do this using the Command window only because doing so is much simpler. If you recall from earlier, the command brought up by clicking on the **Data Editor (Browse)** button was `browse`. We would like to browse only those observations for which `rep78` is missing, so we could type

```
. browse if missing(rep78)
```

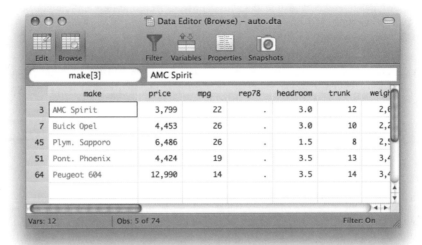

From this, we see that the . entries are indeed missing values—though other missing values are allowable. See [U] **12.2.1 Missing values**. Close the Data Editor after you are satisfied with this statement.

Syntax note: Using the if qualifier above is what allowed us to look at a subset of the observations.

Looking through the data lends no clues about why these particular data are missing. We decide to check the source of the data to see if the missing values were originally missing or if they were omitted in error. Listing the makes of the cars whose repair records are missing will be all we need because we saw earlier that the values of make are unique. This can be done with the menus and a dialog:

1. Select **Data > Describe data > List data**.
2. Click on the drop triangle to the right of the *Variables* field to show the variable names.
3. Click on make to enter it into the *Variables* field.
4. Click on the **by/if/in** tab in the dialog.
5. Type missing(rep78) into the *If: (expression)* box.
6. Click on **Submit**. Stata executes the proper command but the dialog remains open. **Submit** is useful when experimenting, exploring, or building complex commands. We will primarily use **Submit** in the examples. You may click on **OK** in its place if you like.

The same ends could be achieved by typing list make if missing(rep78) in the Command window. The latter is easier, once you know that the command list is used for listing observations. In any case, here is the output:

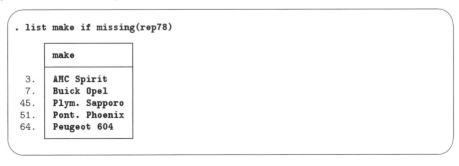

```
. list make if missing(rep78)

         make

  3.   AMC Spirit
  7.   Buick Opel
 45.   Plym. Sapporo
 51.   Pont. Phoenix
 64.   Peugeot 604
```

We go to the original reference and find that the data were truly missing and cannot be resurrected. See [GSM] **10 Listing data and basic command syntax** for more information about all that can be done with the `list` command.

Syntax note: This command uses two new concepts for Stata commands—the `if` qualifier and the `missing()` function. The `if` qualifier restricts the observations on which the command runs to only those observations for which the expression is true. See [U] **11.1.3 if exp**. The `missing()` function tests each observation to see if it contains a missing value. See [U] **13.3 Functions**.

Now that we have a good idea about the underlying dataset, we can investigate the data themselves.

Descriptive statistics

We saw above that the `summarize` command gave brief summary statistics about all the variables. Suppose now that we became interested in the prices while summarizing the data because they seemed fantastically low (it was 1978, after all). To get an in-depth look at the `price` variable, we can use the menus and a dialog:

1. Select **Statistics > Summaries, tables, and tests > Summary and descriptive statistics > Summary statistics**.
2. Enter or select `price` in the *Variables* field.
3. Select *Display additional statistics*.
4. Click on **Submit**.

Syntax note: As can be seen from the Results window, typing `summarize price, detail` will get the same result. The portion after the comma contains *options* for Stata commands; hence, `detail` is an example of an option.

```
. summarize price, detail

                            Price

          Percentiles      Smallest
   1%         3291            3291
   5%         3748            3299
  10%         3895            3667         Obs                  74
  25%         4195            3748         Sum of Wgt.          74

  50%        5006.5                        Mean           6165.257
                             Largest       Std. Dev.      2949.496
  75%         6342           13466
  90%        11385           13594         Variance        8699526
  95%        13466           14500         Skewness       1.653434
  99%        15906           15906         Kurtosis       4.819188
```

From the output, we can see that the median price of the cars in the dataset is only about $5,006! We can also see that the four most expensive cars are all priced between $13,400 and $16,000. If we wished to browse the cars that are the most expensive (and gain some experience with features of the Data Editor), we could start by clicking on the **Data Editor (Browse)** button, ▦. Once the Data Editor is open, we can click the **Filter Observations** button, ▼, to bring up the *Filter Observations* dialog. We can look at the expensive cars by putting `price > 13000` in the *Filter by expression* field:

Pressing the **Apply Filter** button filters the data, and we can see that the expensive cars are two Cadillacs and two Lincolns, which were not designed for gas mileage:

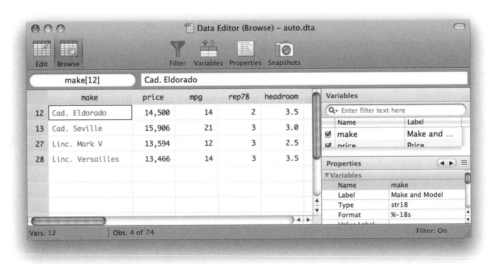

We now decide to turn our attention to foreign cars and repairs because as we glanced through the data, it appeared that the foreign cars had better repair records. (We do not know exactly what the categories 1, 2, 3, 4, and 5 mean, but we know the Chevy Monza was known for breaking down.) Let's start by looking at the proportion of foreign cars in the dataset along with the proportion of cars with each type of repair record. We can do this with one-way tables. The table for foreign cars can be done with menus and a dialog starting with **Statistics > Summaries, tables, and tests > Tables > One-way tables** and then choosing the variable foreign in the *Categorical variable* field. Clicking on **Submit** yields

```
. tabulate foreign
    Car type |      Freq.     Percent        Cum.

    Domestic |         52       70.27       70.27
     Foreign |         22       29.73      100.00

       Total |         74      100.00
```

We see that roughly 70% of the cars in the dataset are domestic, whereas 30% are foreign made. The value labels are used to make the table so that the output is nicely readable.

Syntax note: We also see that this one-way table could be made using the `tabulate` command together with one variable, `foreign`. Making a one-way table for the repair records is simple—it will be simpler if done with the Command window. Typing `tabulate rep78` yields

```
. tabulate rep78
      Repair |
 Record 1978 |      Freq.     Percent        Cum.

           1 |          2        2.90        2.90
           2 |          8       11.59       14.49
           3 |         30       43.48       57.97
           4 |         18       26.09       84.06
           5 |         11       15.94      100.00

       Total |         69      100.00
```

We can see that most cars have repair records of 3 and above, though the lack of value labels makes us unsure what a "3" means. Take our word for it that 1 means a poor repair record and 5 means a good repair record. The five missing values are indirectly evident because the total number of observations listed is 69 rather than 74.

These two one-way tables do not help us compare the repair records of foreign and domestic cars. A two-way table would help greatly, which we can get by using the menus and a dialog:

1. Select **Statistics > Summaries, tables, and tests > Tables > Two-way tables with measures of association**.
2. Choose `rep78` as the *Row variable*.
3. Choose `foreign` as the *Column variable*.
4. It would be nice to have the percentages within the `foreign` variable, so check the *Within-row relative frequencies* checkbox.
5. Click on **Submit**.

Here is the resulting output:

```
. tabulate rep78 foreign, row
```

Key

frequency
row percentage

Repair Record 1978	Car type Domestic	Foreign	Total
1	2	0	2
	100.00	0.00	100.00
2	8	0	8
	100.00	0.00	100.00
3	27	3	30
	90.00	10.00	100.00
4	9	9	18
	50.00	50.00	100.00
5	2	9	11
	18.18	81.82	100.00
Total	48	21	69
	69.57	30.43	100.00

The output indicates that foreign cars are generally much better than domestic cars when it comes to repairs. If you like, you could repeat the previous dialog and try some of the hypothesis tests available from the dialog. We will abstain.

Syntax note: We see that typing the command `tabulate rep78 foreign, row` would have given us the same table. Thus using `tabulate` with two variables yields a two-way table. It makes sense that `row` is an option—we went out of our way to check it in the dialog. Using the `row` option allows us to change the behavior of the `tabulate` command from its default.

Continuing our exploratory tour of the data, we would like to compare gas mileages between foreign and domestic cars, starting by looking at the summary statistics for each group by itself. A direct way to do this would be to use `if` qualifiers to summarize `mpg` for each of the two values of `foreign` separately:

```
. summarize mpg if foreign==0
```

Variable	Obs	Mean	Std. Dev.	Min	Max
mpg	52	19.82692	4.743297	12	34

```
. summarize mpg if foreign==1
```

Variable	Obs	Mean	Std. Dev.	Min	Max
mpg	22	24.77273	6.611187	14	41

It appears that foreign cars get somewhat better gas mileage—we will test this soon.

Syntax note: We needed to use a double equal sign (==) for testing equality. The double equal sign could be familiar to you if you have programmed before. If it is unfamiliar, be aware that it is a common source of errors when initially using Stata. Thinking of equality as "really equal" can cut down on typing errors.

There are two other methods that we could have used to produce these summary statistics. These methods are worth knowing because they are less error-prone. The first method duplicates the concept of what we just did by exploiting Stata's ability to run a command on each of a series of nonoverlapping subsets of the dataset. To use the menus and a dialog, do the following:

1. Select **Statistics > Summaries, tables, and tests > Summary and descriptive statistics > Summary statistics** and click on the **Reset** button, Ⓡ.

2. Select mpg in the *Variables* field.

3. Select the *Standard display* option (if it is not already selected).

4. Click on the **by/if/in** tab.

5. Check the *Repeat command by groups* checkbox.

6. Select or type foreign in the *Variables that define groups* field.

7. **Submit** the command.

You can see that the results match those from above. They have a better appearance because the value labels are used rather than the numerical values. The method is more appealing because the results were produced without knowing the possible values of the grouping variable ahead of time.

```
. by foreign, sort : summarize mpg

-> foreign = Domestic
    Variable |      Obs        Mean    Std. Dev.       Min        Max

         mpg |       52    19.82692     4.743297        12         34

-> foreign = Foreign
    Variable |      Obs        Mean    Std. Dev.       Min        Max

         mpg |       22    24.77273     6.611187        14         41
```

There is something different about the equivalent command that appears above: it contains a *prefix command* called a by prefix. The by prefix has its own option, namely, sort, to ensure that like members are adjacent to each other before being summarized. The by prefix command is important for understanding data manipulation and working with subpopulations within Stata. Make good note of this example, and consult [U] **11.1.2 by varlist:** and [U] **27.2 The by construct** for more information. Stata has other prefix commands for specialized treatment of commands, as explained in [U] **11.1.10 Prefix commands**.

The third method for tabulating the differences in gas mileage across the cars' origins involves thinking about the structure of desired output. We need a one-way table of automobile types (foreign versus domestic) within which we see information about gas mileages. Looking through the menus yields the menu item **Statistics > Summaries, tables, and tests > Tables > One/two-way table of summary statistics**. Selecting this, entering foreign for *Variable 1* and mpg for the *Summarize variable*, and submitting the command yields a nice table:

```
. tabulate foreign, summarize(mpg)
                     Summary of Mileage (mpg)
     Car type         Mean    Std. Dev.         Freq.

     Domestic      19.826923   4.7432972            52
      Foreign      24.772727   6.6111869            22

        Total      21.297297   5.7855032            74
```

The equivalent command is evidently `tabulate foreign, summarize(mpg)`.

Syntax note: This is a one-way table, so `tabulate` uses one variable. The variable being summarized is passed to the `tabulate` command with an option. Though we will not do it here, the `summarize()` option can also be used with two-way tables.

A simple hypothesis test

We would like to run a hypothesis test for the difference in the mean gas mileages. Under the menus, **Statistics > Summaries, tables, and tests > Classical tests of hypotheses > Two-group mean-comparison test** leads to the proper dialog. Enter mpg for the *Variable name* and foreign for the *Group variable name*, and **Submit** the dialog. The results are

```
. ttest mpg, by(foreign)
Two-sample t test with equal variances

   Group       Obs        Mean     Std. Err.   Std. Dev.   [95% Conf. Interval]

Domestic        52      19.82692    .657777    4.743297    18.50638    21.14747
 Foreign        22      24.77273   1.40951     6.611187    21.84149    27.70396

combined        74      21.2973    .6725511    5.785503    19.9569     22.63769

    diff               -4.945804   1.362162                -7.661225   -2.230384

     diff = mean(Domestic) - mean(Foreign)                  t =  -3.6308
Ho: diff = 0                                degrees of freedom =       72

  Ha: diff < 0                 Ha: diff != 0                    Ha: diff > 0
Pr(T < t) = 0.0003     Pr(|T| > |t|) = 0.0005              Pr(T > t) = 0.9997
```

From this, we could conclude that the mean gas mileage for foreign cars is different from that of domestic cars (though we really ought to have wanted to test this before snooping through the data). We can also conclude that the command, `ttest mpg, by(foreign)` is easy enough to remember. Feel free to experiment with unequal variances, various approximations to the number of degrees of freedom, and the like.

Syntax note: The `by()` option used here is not the same as the `by` prefix command used earlier. Although it has a similar conceptual meaning, its usage is different because it is a particular option for the `ttest` command.

Descriptive statistics—correlation matrices

We now change our focus from exploring categorical relationships to exploring numerical relationships: we would like to know if there is a correlation between miles per gallon and weight. We select **Statistics > Summaries, tables, and tests > Summary and descriptive statistics > Correlations and covariances** in the menus. Entering mpg and weight, either by clicking or by typing, and then submitting the command yields

```
. correlate mpg weight
(obs=74)

                |      mpg    weight
     -----------+------------------
            mpg |   1.0000
         weight |  -0.8072    1.0000
```

The equivalent command for this is natural: correlate mpg weight. There is a negative correlation, which is not surprising because heavier cars should be harder to push about.

We could see how the correlation compares for foreign and domestic cars by using our knowledge of how the by prefix works. We can reuse the *correlate* dialog or use the menus as before if the dialog is closed. Click on the **by/if/in** tab, check the *Repeat command by groups* checkbox, and enter the foreign variable to define the groups. As done above on page 12, a simple by foreign, sort: prefix in front of our previous command would work, too:

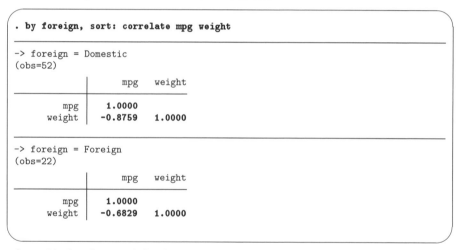

```
. by foreign, sort: correlate mpg weight

-> foreign = Domestic
(obs=52)

                |      mpg    weight
     -----------+------------------
            mpg |   1.0000
         weight |  -0.8759    1.0000

-> foreign = Foreign
(obs=22)

                |      mpg    weight
     -----------+------------------
            mpg |   1.0000
         weight |  -0.6829    1.0000
```

We see from this that the correlation is not as strong among the foreign cars.

Syntax note: Although we used the correlate command to look at the correlation of two variables, Stata can make correlation matrices for an arbitrary number of variables:

```
. correlate mpg weight length turn displacement
(obs=74)

                 |     mpg   weight   length     turn displa~t
    -------------+---------------------------------------------
             mpg |  1.0000
          weight | -0.8072   1.0000
          length | -0.7958   0.9460   1.0000
            turn | -0.7192   0.8574   0.8643   1.0000
    displacement | -0.7056   0.8949   0.8351   0.7768   1.0000
```

This can be useful, for example, when investigating collinearity among predictor variables. In fact, simply typing correlate will yield the complete correlation matrix.

Graphing data

We have found several things in our investigations so far: We know that the average MPG of domestic and foreign cars differs. We have learned that domestic and foreign cars differ in other ways, as well, such as in frequency-of-repair record. We found a negative correlation of MPG and weight—as we would expect—but the correlation appears stronger for domestic cars.

We would now like to examine, with an eye toward modeling, the relationship between MPG and weight, starting with a graph. We can start with a scatterplot of mpg against weight. The command for this is simple: scatter mpg weight. Using the menus requires a few steps because the graphs in Stata may be customized heavily.

1. Select **Graphics > Twoway graph (scatter, line, etc.)**.
2. Click on the **Create...** button.
3. Select the *Basic plots* radio button (if it is not already selected).
4. Select *Scatter* as the basic plot type (if it is not already selected).
5. Select mpg as the *Y variable* and weight as the *X variable*.
6. Click on the **Submit** button.

The Results window shows the command that was issued from the menu:

```
. twoway (scatter mpg weight)
```

The command issued when the dialog was submitted is a bit more complex than the command suggested above. There is good reason for this: the more complex structure allows combining and overlaying graphs, as we will soon see. In any case, the graph that appears is

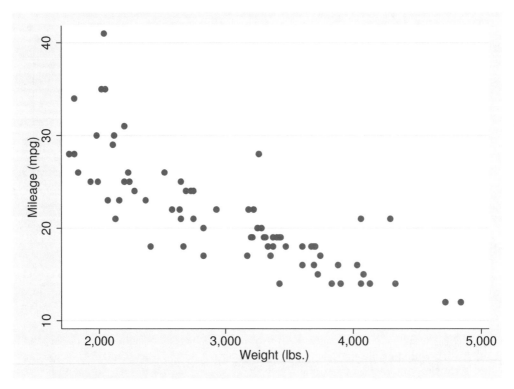

We see the negative correlation in the graph, though the relationship appears to be nonlinear.

Note: When you draw a graph, the Graph window appears, probably covering up your Results window. Click on the **Results** button to put your Results window back on top. Want to see the graph again? Click on the **Graph** button, 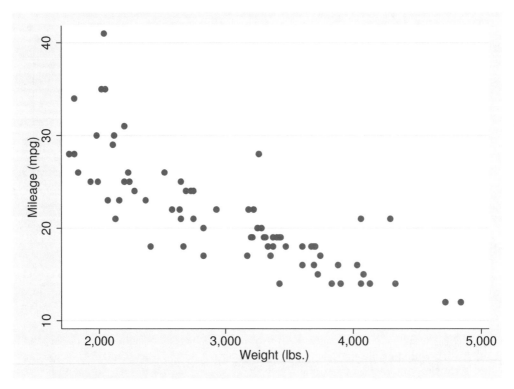. See *The Graph button* in [GSM] **14 Graphing data** for more information about the **Graph** button.

We would now like to see how the different correlations for foreign and domestic cars are manifested in scatterplots. It would be nice to see a scatterplot for each type of car, along with a scatterplot for all the data.

Syntax note: Because we are looking at subgroups, this looks like it is a job for the by prefix. Let's see if this is what we really should use.

Start as before:

1. Select **Graphics > Twoway graph (scatter, line, etc.)** from the menus.
2. If the sheet you used to define the scatterplot is still visible, click on the **Accept** button and skip to step 4.
3. Go through the process on the previous page to create the graph.
4. Click on the **By** tab of the *twoway - Twoway graphs* dialog.
5. Check the *Draw subgraphs for unique values of variables* checkbox.
6. Enter `foreign` in the *Variables* field.
7. Check the *Add a graph with totals* checkbox.
8. Click on the **Submit** button.

The command and the associated graph are

```
. twoway (scatter mpg weight), by(foreign, total)
```

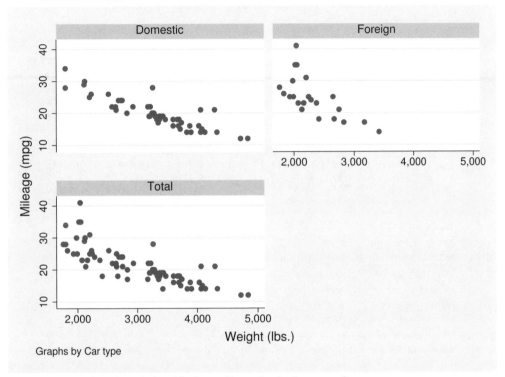

The graphs show that the relationship is nonlinear for both types of cars.

Syntax note: To make the graphs for the combined subgroups, we ended up using a `by()` option, not a `by` prefix. If we had used a `by` prefix, separate graphs would have been generated instead of the combined graph created by the `by` option.

Model fitting: Linear regression

After looking at the graphs, we would like to fit a regression model that predicts MPG from the weight and type of the car. From the graphs, we see that the relationship is nonlinear and so we will try modeling MPG as a quadratic in weight. Also from the graphs, we judge the relationship to be different for domestic and foreign cars. We will include an indicator (dummy) variable for `foreign` and evaluate afterward whether this adequately describes the difference. Thus we will fit the model

$$\text{mpg} = \beta_0 + \beta_1 \, \text{weight} + \beta_2 \, \text{weight}^2 + \beta_3 \, \text{foreign} + \epsilon$$

`foreign` is already an indicator (0/1) variable, but we need to create the weight-squared variable. This can be done with the menus, but here using the command line is simpler. Type

```
. generate wtsq = weight^2
```

Now that we have all the variables we need, we can run a linear regression. We will use the menus and see that the command is also simple. To use the menus, select **Statistics > Linear models and related > Linear regression**. In the resulting dialog, choose mpg as the *Dependent variable* and weight, wtsq, and foreign as the *Independent variables*. **Submit** the command. Here is the equivalent simple regress command and the resulting analysis-of-variance table.

```
. regress mpg weight wtsq foreign

      Source |       SS       df       MS              Number of obs =      74
-------------+------------------------------           F(  3,    70) =   52.25
       Model |  1689.15372        3   563.05124         Prob > F      =  0.0000
    Residual |   754.30574       70  10.7757963         R-squared     =  0.6913
-------------+------------------------------           Adj R-squared =  0.6781
       Total |  2443.45946       73  33.4720474         Root MSE      =  3.2827

-------------------------------------------------------------------------------
         mpg |      Coef.   Std. Err.      t    P>|t|     [95% Conf. Interval]
-------------+-----------------------------------------------------------------
      weight |  -.0165729   .0039692    -4.18   0.000    -.0244892   -.0086567
        wtsq |   1.59e-06   6.25e-07     2.55   0.013     3.45e-07    2.84e-06
     foreign |    -2.2035   1.059246    -2.08   0.041      -4.3161   -.0909002
       _cons |   56.53884   6.197383     9.12   0.000     44.17855    68.89913
-------------------------------------------------------------------------------
```

The results look encouraging, so we will plot the predicted values on top of the scatterplots for each of the types of cars. To do this, we need the predicted, or fitted, values. This can be done with the menus, but doing it by hand is simple enough. We will create a new variable, mpghat. Type

```
. predict mpghat
(option xb assumed; fitted values)
```

The output from this command is simply a notification. Go over to the Variables window and scroll to the bottom to confirm that there is now an mpghat variable. If you were to try this command when mpghat already existed, Stata would refuse to overwrite your data:

```
. predict mpghat
mpghat already defined
r(110);
```

The predict command, when used after a regression, is called a *postestimation command*. As specified, it creates a new variable called mpghat equal to

$$-0.0165729 \, \texttt{weight} + 1.59 \times 10^{-6} \, \texttt{wtsq} - 2.2035 \, \texttt{foreign} + 56.53884$$

For careful model fitting, there are several features available to you after estimation—one is calculating predicted values. Be sure to read [U] **20 Estimation and postestimation commands**.

We can now graph the data and the predicted curve to evaluate separately the fit on the foreign and domestic data to determine if our shift parameter is adequate. We can draw both graphs together. Using the menus and a dialog, do the following:

1. Select **Graphics > Twoway graph (scatter, line, etc.)** from the menus.

2. If there are any plots listed, click on the **Reset** button, Ⓡ.

3. Create the graph for `mpg` versus `weight`:
   a. Click on the **Create...** button.
   b. Be sure that *Basic plots* and *Scatter* are selected.
   c. Select `mpg` as the *Y variable* and `weight` as the *X variable*.
   d. Click on **Accept**.
4. Create the graph showing `mpghat` versus `weight`
   a. Click on the **Create...** button.
   b. Select *Basic plots* and *Line*.
   c. Select `mpghat` as the *Y variable* and `weight` as the *X variable*.
   d. Check the *Sort on x variable* box. Doing so ensures that the lines connect from smallest to largest `weight` values, instead of the order in which the data happen to be.
   e. Click on **Accept**.
5. Show two plots, one each for domestic and foreign cars, on the same graph:
   a. Click on the **By** tab.
   b. Check the *Draw subgraphs for unique values of variables* checkbox.
   c. Enter `foreign` in the *Variables* field.
6. Click on the **Submit** button.

Here are the resulting command and graph:

```
. twoway (scatter mpg weight) (line mpghat weight, sort), by(foreign)
```

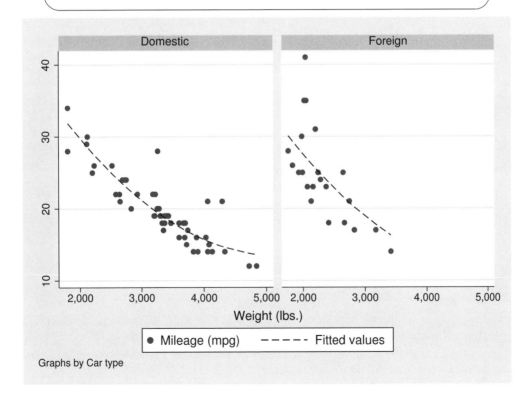

Here we can see the reason for enclosing the separate `scatter` and `line` commands in parentheses: they can thereby be overlaid by submitting them together. The fit of the plots looks good and is cause for initial excitement. So much excitement, in fact, that we decide to print the graph and show it to an engineering friend. We print the graph, being careful to print the graph (and not all our results) by choosing **File > Print > Graph (Graph)**.

When we show our graph to our engineering friend, she is concerned. "No," she says. "It should take twice as much energy to move 2,000 pounds 1 mile compared with moving 1,000 pounds the same distance, and therefore it should consume twice as much gasoline. Miles per gallon is not a quadratic in weight; gallons per mile is a linear function of weight. Don't you remember any physics?"

We try out what she says. We need to generate a gallons-per-mile variable and make a scatterplot. Here are the commands that we would need—note their similarity to commands issued earlier in the session. There is one new command, the `label variable` command, which allows us to give the gpm variable a *variable label* so that the graph is labeled nicely.

```
. generate gp100m = 100/mpg
. label variable gp100m "Gallons per 100 miles"
. twoway (scatter gp100m weight), by(foreign, total)
```

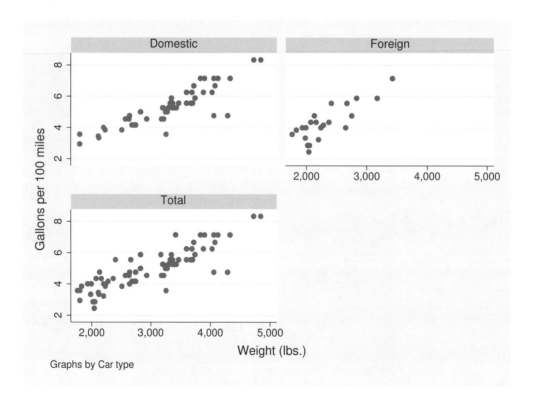

Sadly satisfied that the engineer is indeed correct, we rerun the regression:

```
. regress gp100m weight foreign

      Source |       SS       df       MS              Number of obs =      74
-------------+------------------------------           F(  2,    71) =  113.97
       Model | 91.1761694      2  45.5880847           Prob > F      =  0.0000
    Residual | 28.4000913     71  .400001287           R-squared     =  0.7625
-------------+------------------------------           Adj R-squared =  0.7558
       Total | 119.576261     73  1.63803097           Root MSE      =  .63246

------------------------------------------------------------------------------
      gp100m |      Coef.   Std. Err.      t    P>|t|     [95% Conf. Interval]
-------------+----------------------------------------------------------------
      weight |   .0016254   .0001183    13.74   0.000     .0013896    .0018612
     foreign |   .6220535   .1997381     3.11   0.003     .2237871    1.02032
       _cons |  -.0734839   .4019932    -0.18   0.855    -.8750354    .7280677
------------------------------------------------------------------------------
```

We find that foreign cars had better gas mileage than domestic cars in 1978 because they were so light. According to our model, a foreign car with the same weight as a domestic car would use an additional 5/8 gallon (or 5 pints) of gasoline per 100 miles driven. With this conclusion, we are satisfied with our analysis.

Commands versus menus

In this chapter, you have seen that Stata can operate either with menu choices and dialogs or with the Command window. As you become more familiar with Stata, you will find that the Command window is typically much faster for oft-used commands, whereas the menus and dialogs are faster when building up complex commands, such as those that create graphs.

One of Stata's great strengths is the consistency of its *command syntax*. Most of Stata's commands share the following syntax, where square brackets mean that something is optional and a *varlist* is a list of variables.

$$[prefix:\ ]\ command\ [varlist]\ [if]\ [in]\ [weight]\ [,\ options]$$

Some general rules:

- Most commands accept prefix commands that modify their behavior; see [U] **11.1.10 Prefix commands** for details. One of the more common prefix commands is by.
- If an optional *varlist* is not specified, all the variables are used.
- *if* and *in* restrict the observations on which the command is run.
- *options* modify what the command does.
- Each command's syntax is found in the online help and the reference manuals.
- Stata's command syntax includes more than we have shown you here, but this introduction should get you started. For more information, see [U] **11 Language syntax** and help language.

We saw examples using all the pieces of this except for the in qualifier and the weight clause. The syntax for all commands can be found in the online help along with examples—see [GSM] **4 Getting help** for more information. The consistent syntax makes it straightforward to learn new commands and to read others' commands when examining an analysis.

Here is an example of reading the syntax diagram that uses the summarize command from earlier in this chapter. The syntax diagram for summarize is typical:

$$summarize\ [varlist]\ [if]\ [in]\ [weight]\ [,\ options]$$

This means that

command by itself is valid: summarize

command followed by a varlist
(variable list) is valid: summarize mpg
                                                summarize mpg weight

command with if (with or without
a varlist) is valid: summarize if mpg>20
                                                summarize mpg weight if mpg>20

and so on.

You can learn about summarize in [R] **summarize**, or select **Help > Stata Command...** and enter summarize, or type help summarize in the Command window.

Keeping track of your work

It would have been useful if we had made a log of what we did so that we could conveniently look back at interesting results or track any changes that were made. You will learn to do this in [GSM] **16 Saving and printing results by using logs**. Your logs will contain commands and their output—another reason to learn command syntax, so that you can remember what you have done.

To make a log file that keeps track of everything appearing in the Results window, click on the **Log** button, that looks like a lab notebook, . Choose a place to store your log file and give it a name, just as you would any other document. The log file will save everything that appears in the Results window from the time you start a log file to the time that you close it.

Conclusion

This chapter introduced you to Stata's capabilities. You should now read and work through the rest of this manual. Once you are done here, you can read the *User's Guide.*

2 The Stata user interface

The windows

This chapter introduces the core of Stata's interface: its main windows, its toolbar, its menus, and its dialogs.

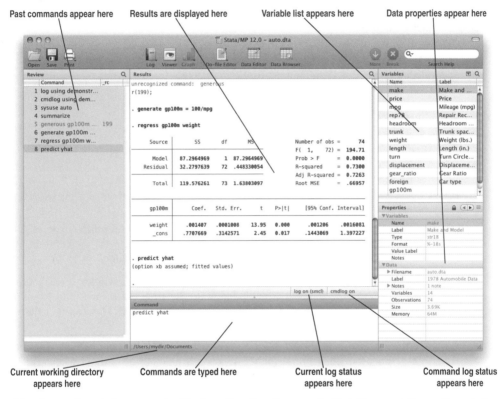

Past commands appear here   Results are displayed here   Variable list appears here   Data properties appear here

Current working directory appears here   Commands are typed here   Current log status appears here   Command log status appears here

The five main windows are the Review, Results, Command, Variables, and Properties windows. Each window has its name in its title bar. These five windows are typically in use the whole time Stata is open. There are other, more specialized windows such as the Viewer, Data Editor, Variables Manager, Do-file Editor, Graph, and Graph Editor windows—these are discussed later in this manual.

To open any window or to reveal a window hidden by other windows, select the window from the **Window** menu or select the proper item from the toolbar. You can also use Exposé in Mac OS X 10.5 or later to reveal hidden windows (*F9* is best), or use *Command–`* (left quote) to cycle through all open Stata windows. Many of Stata's windows have functionality that can be accessed by clicking on the right mouse button (right-clicking) within the window. If your mouse has only one button, press the *Control* key while pressing the mouse button to simulate a right-click. Right-clicking displays a contextual menu that, depending on the window, allows you to copy text, set the preferences for the window, or print the contents of the window. When copying text or printing, we recommend that you always right-click on the window rather than use the menu bar or toolbar so that you can be sure of where and what you are copying or printing.

The toolbar

This is the toolbar:

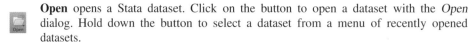

The toolbar contains buttons that provide quick access to Stata's more commonly used features. If you forget what a button does, hold the mouse pointer over the button for a moment, and a tooltip will appear with a description of that button.

Buttons that include both an icon and an arrow display a menu if you hold down the mouse button. Here is an overview of the toolbar buttons and their functions:

Open opens a Stata dataset. Click on the button to open a dataset with the *Open* dialog. Hold down the button to select a dataset from a menu of recently opened datasets.

Save saves the Stata dataset currently in memory to disk.

Print displays a list of windows. Select a window name to print its contents.

Log begins a new log or closes, suspends, or resumes the current log. See [GSM] **16 Saving and printing results by using logs** for an explanation of log files. You can also choose a file to view in the Viewer.

Viewer opens the Viewer or brings a Viewer to the front of all other windows. Click on the button to open a new Viewertab.Click and hold to select a Viewer to bring to the front. See [GSM] **3 Using the Viewer** for more information.

Graph brings the Graph window to the front of all other windows. Click on the button to bring the topmost Graph window to the front. Click and hold to select a graph to bring to the front. See *The Graph button* in [GSM] **14 Graphing data** for more information.

Do-file Editor opens the Do-file Editor or brings a Do-file Editor to the front of all other windows. Click on the button to open a new Do-file Editor. Click and hold to select a Do-file Editor to bring to the front. See [GSM] **13 Using the Do-file Editor—automating Stata** for more information.

Data Editor opens the Data Editor or brings the Data Editor to the front of the other Stata windows. See [GSM] **6 Using the Data Editor** for more information.

Data Browser opens the Data Editor in browse mode. See *Browse mode* in [GSM] **6 Using the Data Editor** for more information.

More tells Stata to continue when it has paused in the middle of long output. See [GSM] **10 Listing data and basic command syntax** for more information.

Break stops the current task in Stata. See [GSM] **10 Listing data and basic command syntax** for more information.

The Command window

Commands are submitted to Stata from the Command window. The Command window supports basic text editing, copying and pasting, a command history, function-key mapping, and variable-name completion. The Command window also uses the same syntax highlighting as the Do-file Editor. See [GSM] **13 Using the Do-file Editor—automating Stata** for more information.

From the Command window, pressing

Page Up	steps backward through the command history.
Page Down	steps forward through the command history.
Tab	auto-completes a partially typed variable name, when possible.

See [U] **10 Keyboard use** for more information about keyboard shortcuts for the Command window.

The command history allows you to recall a previously submitted command, edit it if you wish, and then resubmit it. Commands submitted by Stata's dialogs are also included in the command history, so you can recall and submit a command without having to open the dialog again.

The Results window

The Results window contains all the commands and their textual results you have entered during the Stata session. While you can scroll through the Results window to look at work you have done, it is much simpler to search within the Results window using the find bar. By default, the find bar is hidden. You can toggle its visibility by clicking the magnifying glass button, Q , in the Review window titlebar.

The Review window

The Review window shows the history of commands that have been entered. It displays successful commands in black and unsuccessful commands, along with their error codes, in red.

The toolbar has two tools for manipulating the contents of the Review window. Clicking on the magnifying glass button, Q , in the Review window titlebar toggles the visibility of these tools. Text entered in the entry field will filter the commands appearing in the Review window. By default, the filter will ignore case and find any commands containing any of the words in the filter. Clicking on the arrow by the magnifying glass will allow you to change this behavior. Clicking on the **Hide Errors** button toggles the hiding of commands that ran with an error.

No commands are deleted by using these tools—all that is affected is their visibility.

To enter a command from the Review window, you can

- Click once on a past command to copy it to the Command window, replacing the contents of the Command window.
- Double-click on a past command to execute it. Executing the command adds the command to the bottom of the Review window, also.

Right-clicking on the Review window displays a menu from which you can select various actions:

- **Cut** removes the selected commands from the Review window and places them on the Clipboard.
- **Copy** copies the selected commands to the Clipboard.
- **Delete** removes the selected commands from the Review window.
- **Select All** selects all the commands in the Review window, including those before and after the commands currently displayed.
- **Clear All** clears out all the commands from the Review window, including those before and after the commands currently displayed.
- **Do Selected** submits all the selected commands and adds them to the bottom of the command history. Stata will attempt to run all the selected commands, even those containing errors, and will not stop even if a command causes an error.

- **Send to Do-file Editor** places all the selected commands into a new Do-file Editor window.
- **Save All...** brings up a *Save Review Contents* dialog, which allows you to save all the commands in the Review window, including those before and after the commands currently displayed, in a *do-file*. (See [GSM] **13 Using the Do-file Editor—automating Stata** for more information on do-files.)
- **Save Selected...** brings up a *Save Review Contents* dialog, which allows you to save the selected commands in the Review window in a do-file.
- **Preferences...** allows you to edit the preferences for the Review window.

The Variables window

The Variables window shows the list of variables in the dataset, along with selected properties of the variables. By default, it shows all the variables and their variable labels. You can change what properties get displayed by right-clicking on the header of any column of the Variables window.

Click once on a variable in the Variables window to select it. Multiple variables can be selected in the usual fashion, either by *Command*-clicking on nonadjacent variables or by clicking on a variable and *shift*-clicking on a second variable to select all intervening variables.

Double-clicking on a variable in the Variables window puts the selected variable at the insertion point in the Command window. If multiple variables are selected, double-clicking on any selected variable will put all the selected variables at the insertion point in the Command window.

The leftmost column of the Variables window is called the one-click paste column. You can also send variables to the Command window by hovering the mouse over the one-click paste column of the Variables window and clicking on the arrow that appears.

The Variables window supports filtering and reordering of variables. Clicking on the magnifying glass button, Q , in the Variables window titlebar toggles the visibility of the filter. Text entered in the entry field will filter the variables appearing in the Variables window. The filter is applied to all visible columns and shows all variables that match the criteria in at least on column. By default, the filter will ignore case and show any variables for which at least one column contains any of the words in the filter. Clicking on the arrow by the magnifying glass will allow you to change this behavior.

You can reorder the variables in the Variables window by clicking on any column header. The first click sorts in ascending order, the second click sorts in descending order, and the third click puts the variables back in dataset order. Thus clicking on the *Variable* column header will make the Variables window display the variables in alphabetical order. Sorting in the Variables window is live, so if you change a property of a variable when the Variables window is sorted by that property, it will automatically move the variable to its proper location. Reordering the display order of the variables in the Variables window does not affect the order of the variables in the dataset itself.

Right-clicking on a variable in the Variables window displays a menu from which you can select

- **Keep Only Variable** *varname* (or **Keep Only Selected Variables** if multiple variables are selected) to keep just the selected variables in the dataset in memory. You will be asked for confirmation. This affects only the dataset in memory, not the dataset as saved on your disk. See [GSM] **12 Deleting variables and observations** for more information.
- **Drop Variable** *varname* (or **Drop Selected Variables** if multiple variables are selected) to drop, or eliminate, the selected variables from the dataset in memory. You will be asked for confirmation. Just as above, this affects only the dataset in memory, not the dataset as saved on your disk. See [GSM] **12 Deleting variables and observations** for more information.
- **Copy Varlist** to copy the selected variable names to the clipboard.

- **Select All** to select all variables in the dataset that satisfy the filter conditions. If no filter has been specified, all variables will be selected.
- **Send Varlist to Command Window** to send all selected variables to the Command window.
- **Preferences...** to change the preferences for the Variables window.

Items from the contextual menu issue standard Stata commands, so working by right-clicking is just like working directly in the Command window.

If you would like to hide the Variables window, grab the divider between the Variables window and the Results window and drag it all the way to the right. This is like resizing the Variables window to have zero width. Hiding the Variables window will also hide the Properties window.

To reveal a hidden Variables window, drag the right edge of the main Stata window to the left. You can also select **Window > Variables** to expose a narrow Variables window.

The Properties window

The Properties window displays variable and dataset properties. If a single variable is selected in the Variables window, its properties are displayed. If there are multiple variables selected in the Variables window, the Properties window will display properties that are common across all selected variables.

Clicking the lock icon in the Properties window titlebar toggles the ability to alter properties of the selected variable(s). By default, changes are not allowed. Once the properties are unlocked, you can make any changes to variable or dataset properties you like. Each change you make will create a command that appears in the Results and Command windows, as well as in any command log, so the changes are reproducible. Using the Properties window is one of the simplest ways of managing notes, changing variable and value labels, and changing display formats. See [D] **notes**, [D] **label**, and [D] **format**.

Clicking the arrow buttons next to the lock icon will select the previous or next variable shown in the Variables window and that selection will be reflected in the Properties window.

If you would like to hide the Properties window, click on the hide triangle in the titlebar of the Variables window. If you would like to reveal a hidden Properties window, click on the expose triangle in the titlebar of the Variables window.

You should also investigate the Variables Manager, explained in [GSM] **7 Using the Variables Manager**, because it extends these capabilities and provides a good interface for managing variables.

Menus and dialogs

There are two ways by which you can tell Stata what you would like it to do: you can use menus and dialogs, or you can use the Command window. When you worked through the sample session in [GSM] **1 Introducing Stata—sample session**, you saw that both ways have strengths. We will discuss the menus and dialogs here.

Stata's **Data**, **Graphics**, and **Statistics** menus provide point-and-click access to almost every command in Stata. As you will learn, Stata is fully programmable, and Stata programmers can even create their own dialogs and menus. The **User** menu provides a place for programmers to add their own menu items. Initially, it contains only some empty submenus.

If you wish to perform a Poisson regression, for example, you could type Stata's `poisson` command or you could select **Statistics > Count outcomes > Poisson regression**, which would display this dialog:

This dialog provides access to all the functionality of Stata's `poisson` command. The `poisson` command has many options that can be accessed by clicking on the multiple tabs across the top of the dialog. The first time you use the dialog for a command, it is a good idea to look at the contents of each tab so that you will know all the dialog's capabilities.

The dialogs for many commands have the **by/if/in** and **Weights** tabs. These provide access to Stata's commands and qualifiers for controlling the estimation sample and dealing with weighted data. See [U] **11 Language syntax** for more information on these features of Stata's language.

The dialogs for most estimation commands have the **Maximization** tab for setting the maximization options (see [R] **maximize**). For example, you can specify the maximum number of iterations for the optimizer.

Most dialogs in Stata provide the same six buttons you see at the bottom of the *poisson* dialog above.

OK issues a Stata command based on how you have filled out the fields in the dialog, and then closes the dialog.

Cancel closes the dialog without doing anything—just as clicking on the dialog's red close button does.

Submit issues a command just like **OK** but leaves the dialog on the screen so that you can make changes and issue another command. This feature is handy when, for example, you are learning a new command or putting together a complicated graph.

Help provides access to Stata's help system. Clicking on this button will typically take you to the help file for the Stata command associated with the dialog. Clicking on it here would take you to the poisson help file. The help file will have tabs above groups of options to show which dialog tab contains which options.

Reset resets the dialog to its default state. Each time you open a dialog, it will remember how you last filled it out. If you wish to reset its fields to their default values at any time, simply click on this button.

Copy Command to Clipboard behaves much like the **Submit** button, but rather than issuing a command, it copies the command to the Clipboard. The command can then be pasted elsewhere (such as in the Do-file Editor).

The command issued by a dialog is submitted just as if you had typed it by hand. You can see the command in the Results window and in the Review window after it executes. Looking carefully at the full command will help you learn Stata's command syntax.

In addition to being able to access the dialogs for Stata commands through Stata's menus, you can also invoke them by using two other methods. You may know the name of a Stata command for which you want to see a dialog, but you might not remember how to navigate to that command in the menu system. Simply type db *commandname* to launch the dialog for *commandname*:

```
. db poisson
```

You will also find access to the dialog for a command in that command's help file; see [GSM] **4 Getting help** for more details.

As you read this manual, we will present examples of Stata commands. You may type those examples as presented, but you should also experiment with submitting those commands by using their dialogs. Use the db command described above to quickly launch the dialog for any command that you see in this manual.

The working directory

If you look at the screen shot on page 23, you will notice the status bar at the base of the main Stata window that contains the name of the current working directory /Users/mydir/Documents. This path indicates that /Users/mydir/Documents is the current working directory. Each session, this is set to the working directory you were using when you last quit Stata. The current working directory is the folder where graphs and datasets will be saved when typing commands such as save *filename*. It does not affect the behavior of menu-driven file actions such as **File > Save** or **File > Open...**. Once you have started Stata, you can change the current working directory with the cd command. See [D] **cd** for full details. Stata always displays the name of the current working directory so that it is easy to tell where your graphs and datasets will be saved.

Notes

3 Using the Viewer

The Viewer's purpose

The Viewer is a versatile tool in Stata. It will be the first place you can turn for help within Stata, but it is far more than just a help system. You can also use the Viewer to keep your copy of Stata current; to add, delete, and manage third-party extensions to Stata that are known as *user-written programs*; to view and print Stata logs both from your current and previous Stata sessions; to view and print any other Stata-formatted (SMCL) or plain-text (ASCII) file; and even to launch your web browser to follow hyperlinks.

This chapter focuses on the general use of the Viewer, its buttons, and a brief summary of the commands that the Viewer understands. There is more information about using the Viewer to find help in [GSM] **4 Getting help** and for installing user-written commands in [GSM] **19 Updating and extending Stata—Internet functionality**.

To open a new Viewer window, (or open a new tab in an existing Viewer), you may either click on the **Viewer** button, 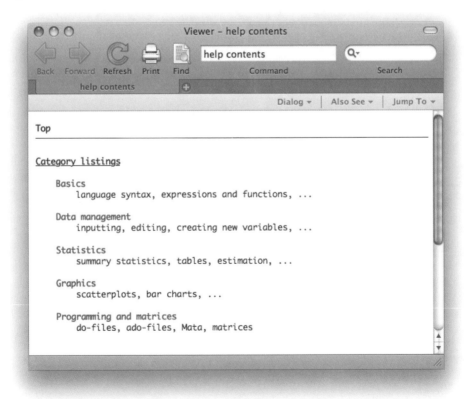, or select **Window > Viewer > New Viewer**. A Viewer opened by these means provides links that allow you to perform several tasks.

Viewer buttons

The toolbar of the Viewer has multiple buttons, a command box, and a search box.

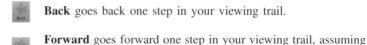

Back goes back one step in your viewing trail.

Forward goes forward one step in your viewing trail, assuming you backtracked.

Refresh refreshes the Viewer, in case you are viewing something that has changed since you opened the Viewer.

Print prints the contents of the Viewer.

Find opens the find bar at the bottom of the Viewer (see below).

Search Chooses the scope of help searches in the Viewer.

The Find bar is used to find text within the current Viewer. To reveal the Find bar at the bottom of the window, click on the **Find** button (see above):

The Find bar has its own buttons, fields, and checkboxes.

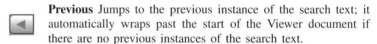

Previous Jumps to the previous instance of the search text; it automatically wraps past the start of the Viewer document if there are no previous instances of the search text.

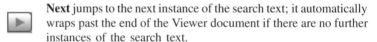

Next jumps to the next instance of the search text; it automatically wraps past the end of the Viewer document if there are no further instances of the search text.

Find is the field for entering the search text you would like to find. You can change the search options by clicking on the down arrow near the magnifying glass.

Done closes the find bar.

Viewer's function

The Viewer is similar to a web browser. It has links (shown in blue text) that you can click on to see related help topics and to install and manage third-party software. When you move the mouse pointer over a link, the status bar at the bottom of the Viewer shows the action associated with that link. If the action of a link is `help logistic`, clicking on that link will show the help file for the `logistic` command in the Viewer. Middle-clicking on a link in a Viewer window (if you do not have a three-button mouse, then *Shift*–clicking) will open the link in a new Viewer window. *Command*–clicking will open the link in a new tab in the current Viewer window.

You can open a new Viewer by selecting **Window > Viewer > New Viewer** or by clicking on the **Viewer** button on the toolbar. Entering a `help` command from the Command window will also open a new Viewer.

To bring a Viewer to the front of all other Viewers, select **Window > Viewer** and choose a Viewer from the list there. Selecting **Close All Viewers** closes all open Viewer windows.

Viewing local text files, including SMCL files

In addition to viewing built-in Stata help files, you can use the Viewer to view Stata Markup and Control Language (SMCL) files such as those typically produced when logging your work (see [GSM] **16 Saving and printing results by using logs**), as well as plain-text files. To open a file and view its contents, simply select **File > View...**, and you will be presented with a dialog:

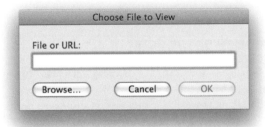

You may either type in the name of the file that you wish to view and click on **OK**, or you may click on the **Browse...** button to open a standard file dialog that allows you to navigate to the file.

If you currently have a log file open, you may view the log file in the Viewer. This method has one advantage over scrolling back in the Results window: what you view stays fixed even as output is added to the Results window. If you wish to view a current log file, select **File > Log > View...**, and the usual dialog will appear, but with the path and filename of the current log already in the field. Simply click on **OK**, and the log will appear in the Viewer. See [GSM] **16 Saving and printing results by using logs** for more details.

Viewing remote files over the Internet

If you want to look at a remote file over the Internet, the process is similar to viewing a local file, only instead of using the **Browse...** button, you type the URL of the file that you want to see, such as http://www.stata.com/man/readme.smcl. You should use the Viewer only to view text or SMCL files. If you enter the URL of, say, an arbitrary webpage, you will see the HTML source of the page instead of the usual browser rendering.

Navigating within the Viewer

In addition to using the window scrollbar to navigate the Viewer window, you also can use the up and down cursor keys and *Page Up* and *Page Down* keys to do the same. Pressing the up/down cursor keys scrolls the window a line at a time. Pressing the *Page Up/Page Down* keys scrolls the window a screenful at a time.

Printing

To print the contents of the Viewer, right-click on the window and select **Print...**. You may also select **File > Print >** *Viewer name* or click and hold the **Print** toolbar button to select from a menu of open windows to print.

Tabs in the Viewer

The Viewer window can have multiple tabs. You may view different files or different views of the same file in different tabs. Clicking on the **Open New Tab** button will open a new tab in the current Viewer window. You can change the order of the tabs within a Viewer by dragging the tabs along the tab bar within the window. If you drag a tab off the tab bar, you will create a new Viewer window to hold the tab. You can also drag tabs from one Viewer window to another. If you drag the last tab from a Viewer into another Viewer, the original Viewer window will close. In this way you can combine Viewer windows.

Right-clicking on the Viewer window

Right-clicking on the Viewer window displays a contextual menu that offers these options:
- **Select All** to select all text in the Viewer.
- **Preferences...** to edit the preferences for the Viewer window.
- **Print...** to print the contents of the Viewer window.

Searching for help in the Viewer

The search box in the Viewer can be used to search documentation. Click on the magnifying glass; choose *Search documentation and FAQs*, *Search net resources*, or *Search all*; and then type a word or phrase in the search box and press *Return*. For more extensive information about using the Viewer for help, see [GSM] **4 Getting help**.

Commands in the Viewer

Everything you can do in the Viewer by clicking on links and buttons can also be done by typing commands in the command box at the top of the window or on the Stata command line. Some tasks that can be performed in the Viewer are
1. Obtaining help (see [GSM] **4 Getting help**)
     Type `contents` to view the contents of Stata's help system.
     Type *commandname* to view the help file for a Stata command.

2. Searching (see [GSM] **4 Getting help**)

> Type `search` *keyword* to search documentation and FAQs on a topic.
>
> Type `search` *keyword*, `net` to search net resources on a topic.
>
> Type `search` *keyword*, `all` to search both of the above.

3. Finding and Installing User-written programs (see [GSM] **4 Getting help** and [GSM] **19 Updating and extending Stata—Internet functionality**)

> Type `net from http://www.stata.com/` to find and install *Stata Journal*, *Stata Technical Bulletin*, and user-written programs from the Internet.
>
> Type `ado` to review user-written programs you have installed.
>
> Type `ado uninstall` to uninstall user-written programs you have installed on your computer.

4. Updating (see *Official Stata updates* in [GSM] **19 Updating and extending Stata—Internet functionality**)

> Type `update` to check your current Stata version.
>
> Type `update query` to check for new official Stata update releases.
>
> Type `update all` to update your Stata.

5. Viewing files in the Viewer

> Type `view` *filename*`.smcl` to view SMCL files.
>
> Type `view` *filename*`.txt` to view text files.
>
> Type `view` *filename*`.log` to view text log files.

6. Viewing files in the Results window

> Type `type` *filename*`.smcl` in the Command window to view SMCL files in the Results window.
>
> Type `type` *filename*`.txt` in the Command window to view text files in the Results window.
>
> Type `type` *filename*`.log` in the Command window to view text log files in the Results window.

7. Launching your browser to view an HTML file

> Type `browse` *URL* to launch your browser.

8. Keeping informed

> Type `news` to see the latest news from http://www.stata.com.

Using the Viewer from the Command window

Typing `help` *commandname* in the Command window will bring up a new Viewer showing the requested help.

Notes

4 Getting help

Online help

Stata's help system provides a wealth of information to help you learn and use Stata. To find out which Stata command will perform the statistical or data-management task you would like to do, you should generally follow these steps:

1. Select **Help > Search...**, choose *Search documentation and FAQs*, and enter the topic or keywords. This search will open a new Viewer window containing information about Stata commands, references to articles in the *Stata Journal* or the *Stata Technical Bulletin* (STB), links to Frequently Asked Questions (FAQs) on Stata's website, and links to selected external websites.

2. Read through the results, and click on the link to the appropriate command name to open its help file.

3. Read the help file for the command you chose.

4. If you want more in-depth help, click on the link from the name of the command to the PDF documentation, read it, then come back to Stata.

5. If the first help file you went to is not what you wanted, either click on the **Also See** button and choose a links to related help files or click on the **Back** button to go back to the previous document and go from there to other help files.

6. With the help file open, click on the Command window and enter the command, or click on the **Dialog** button and choose a link to open a dialog for the command.

7. If, at any time, you want to begin again with a new search, enter the new search terms in the search box of the Viewer window.

8. If your *Search documentation and FAQs* search returned no results, you can look for *Stata Journal*, STB, and user-written programs on the Internet by entering your search term(s) in the search box of the Viewer window, then clicking on the magnifying glass, and selecting *Search net resources* from the resulting pulldown menu.

9. If you select *Search documentation and FAQs*, Stata searches its keyword database. If you select *Search net resources*, Stata searches for *Stata Journal*, STB, and user-written programs available for free download on the Internet; see [GSM] **19 Updating and extending Stata—Internet functionality** for more information.

10. You can also select *Search all*, which is equivalent to choosing both *Search documentation and FAQs* and *Search net resources*. This is the equivalent of Stata's `findit` command.

Let's illustrate the help system with an example. You will get the most benefit from the example if you work along at your computer.

Suppose that we have been given a dataset about antique cars and we need to know what it contains. Though we still have a vague notion of having seen something like this while working through the example session in [GSM] **1 Introducing Stata—sample session**, we do not remember the proper command.

Start by typing `sysuse auto, clear` in the Command window to bring the dataset into memory. (See [GSM] **5 Opening and saving Stata datasets** for information on the `clear` option.)

Following the above approach, we

1. Select **Help > Search...**.

2. Check that the *Search documentation and FAQs* radio button is selected.

3. Type `dataset contents` into the search box, and click on **OK** or press *Return*. Before pressing *Return*, the window should look like

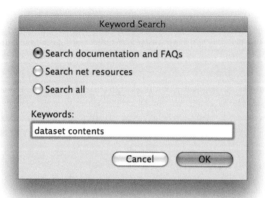

4. Stata will now search for "dataset contents" among the Stata commands, the reference manuals, the *Stata Journal*, the *Stata Technical Bulletin*, and the FAQs on Stata's website. Here is the result:

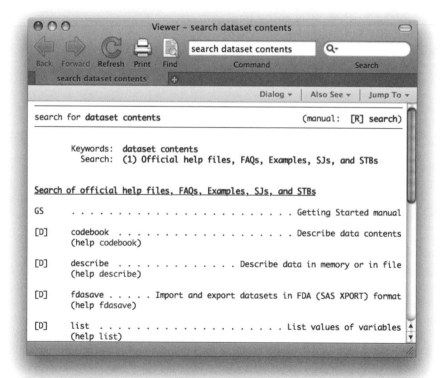

5. Upon seeing the results of the search, we see two commands that look promising: `codebook` and `describe`. Because we are interested in the contents of the dataset, we decide to check out

the codebook command. The [D] means that we could look up the codebook command in the *Data-Management Reference Manual*. The blue codebook link in (help codebook) means that there is an online help file for the codebook command. This is what we are interested in right now.

6. Click on the blue codebook link. Links can take you to a variety of resources, such as help for Stata commands, dialog, and even webpages. Here the link goes to the help file for the codebook command.

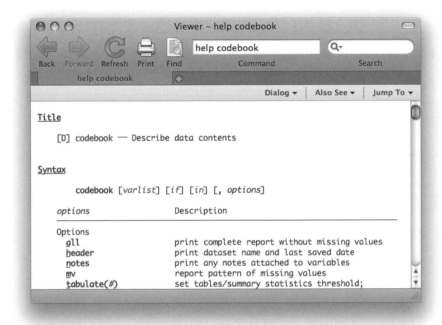

7. What is displayed is typical for help for a Stata command. Help files for Stata commands contain, from top to bottom, these features:

   a. The quick access toolbar with three buttons:

      i. The **Dialog** button shows links to any dialogs associated with the command.

      ii. The **Also See** button shows links to related PDF documentation and help files.

      iii. The **Jump To** button shows links to other sections within the current help file.

   b. A link to the manual entry for the command in the PDF documentation. Clicking on the link will open your PDF viewer and show you the complete documentation for the command—in this case, codebook.

   c. The command's syntax, that is, rules for constructing a command that Stata will correctly interpret. The square brackets here indicate that all the arguments to codebook are optional but that if we wanted to specify them, we could use a *varlist*, an if qualifier, or an in qualifier, along with some options. (Options vary greatly from command to command.) The options are listed directly under the command and are explained in some detail later in the help file. You will learn more about command syntax in [GSM] **10 Listing data and basic command syntax**.

   d. The location in the menu system for the command.

e. A description of the command. Because "codebook" is the name for big binders containing hard copy describing each of the elements of a dataset, the description for the `codebook` command is justifiably terse.

f. The options that can be used with this command. These are explained in much greater detail than in the listing of the possible options after the syntax. Here, for example, we can see that the `mv` option can look to see if there is a pattern in the missing values—something important for data cleaning and imputation.

g. Examples of command usage. The `codebook` examples are real examples that step through using the command on a dataset either shipped with Stata or loadable within Stata from the Internet.

h. The information the command saves in the returned results. These results are used primarily by programmers.

For now, either click on **Jump To** and choose **Examples** from the dropdown menu or scroll down to the examples. It is worth going through the examples as given in the help file. Here is a screen shot of the top of the examples:

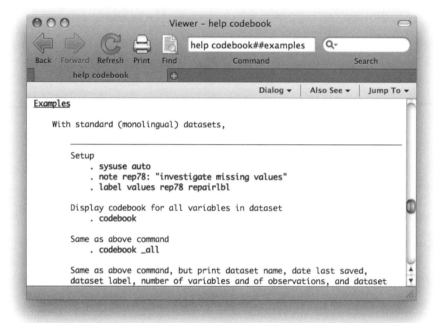

Searching help

Search is designed to help you find information about statistics, graphics, data management, and programming features in Stata. When entering topics for the search, use appropriate terms from statistics, etc. For example, you could enter `Mann-Whitney`. Multiple topic words are allowed, for example, `regression residuals`. For advice on using **Search**, click on the **Advice** button, and follow your choice of advice links.

When you are using **Search**, use proper English and proper statistical terminology. If you already know the name of the Stata command and want to go directly to its help file, select **Help > Stata Command...** and type the command name. You can also type the command name in the *Search* field at the top of the Viewer and press *Return*.

Help distinguishes between topics and Stata commands, because some names of Stata commands are also general topic names. For example, `logistic` is a Stata command. If you choose **Stata Command...** and type `logistic`, you will go right to the help file for the command. But if you choose **Search...** and type `logistic`, you will get search results listing the many Stata commands that relate to logistic regression.

Remember that you can search for help from within a Viewer window by typing a command in the command box of the Viewer or by clicking the magnifying glass button to the left of the search box, selecting the scope of your search, typing the search criteria in the search box, and pressing *Return*.

Help and search commands

As you might expect, the help system is accessible from the Command window. This feature is especially convenient when you need help on a particular Stata command. Here is a short listing of the various commands you can use:

- Typing `search` *topic* in the Command window produces the same output as selecting **Help > Search...**, choosing *Search documentation and FAQs*, and typing *topic*. The output appears in the Results window instead of a Viewer.
- Typing `search` *topic*`, net` in the Command window produces the same output as selecting **Help > Search...**, choosing *Search net resources*, and typing *topic*. The output appears in the Results window instead of a Viewer.
- Typing `search` *topic*`, all` in the Command window produces the same output as selecting **Help > Search...**, choosing *Search all*, and typing *topic*. The output appears in the Results window instead of a Viewer.
- Typing `findit` *topic* in the Command window behaves like `search` *topic*`, all`, except the output appears in a Viewer window. This is often the fastest way to search.
- Typing `help` *commandname* is equivalent to selecting **Help > Stata Command...** and typing *commandname*. The help file for the command appears in a new Viewer window.
- Typing `chelp` *commandname* is similar to selecting **Help > Stata Command...** and typing *commandname*, except the help file for the command appears in the Results window instead of a Viewer window.

See [U] **4 Stata's help and search facilities** and [U] **4.9 search: All the details** in the *User's Guide* for more information about these command-language versions of the help system. The `search` command, in particular, has a few capabilities (such as author searches) that we have not demonstrated here.

The Stata reference manuals and User's Guide

Notations such as [R] **ci**, [R] **regress**, and [R] **ttest** in the **Search** results and help files are references to the four-volume *Base Reference Manual*. You may also see things like [P] **#delimit**, which is a reference to the *Programming Reference Manual*, and [U] **9 The Break key**, which is a reference to the *User's Guide*. For a complete list of manuals and their shorthand notations, see *Cross-referencing the documentation*, which immediately follows the table of contents in this manual. Many of the links in the help files point to the PDF versions of the manuals that came with Stata. It is worth clicking on these links to read the extensive information found in the manuals. The Stata help system, though extensive, contains only a fraction of the information found in the manuals.

The Stata reference manuals are each arranged like an encyclopedia, alphabetically, and each has its own index. The *User's Guide* also has an index. The *Quick Reference and Index* contains a combined index for the *User's Guide* and all the reference manuals. This combined index is a good place to start when you are looking for information about a command.

Entries have names like **collapse**, **egen**, and **summarize**, which are generally themselves Stata commands.

For advice on how to use the reference manuals, see [GSM] **18 Learning more about Stata**, or see [U] **1.1 Getting Started with Stata**.

The Stata Journal and the Stata Technical Bulletin

The *Search documentation and FAQs* facility searches all the Stata source materials, including the online help, the *User's Guide*, the reference manuals, this manual, the *Stata Journal*, the *Stata Technical Bulletin* (STB), and the FAQs on Stata's website.

The *Search net resources* facility searches all materials available via Stata's net command; see [R] **net**. These materials include the *Stata Journal*, STB, and user-written additions to Stata available on the Internet.

The *Stata Journal* is a printed and electronic journal, published quarterly, containing articles about statistics, data analysis, teaching methods, and effective use of Stata's language. The *Journal* publishes reviewed papers together with shorter notes and comments, regular columns, tips, book reviews, and other material of interest to researchers applying statistics in a variety of disciplines. The *Journal* is a publication for all Stata users, both novice and experienced, with different levels of expertise in statistics, research design, data management, graphics, reporting of results, and of Stata, in particular. See http://www.stata-journal.com for more information. There, you may browse the archive, subscribe, order PDF copies of individual articles, and view at no charge copies of articles older than 3 years.

The predecessor to the *Stata Journal* was the *Stata Technical Bulletin* (STB). Even though the STB is no longer published, past issues contain articles and programs that may interest you. See http://www.stata.com/bookstore/stbj.html for the table of contents of past issues, and see the STB FAQs at http://www.stata.com/support/faqs/res/stb.html for detailed information on the STB.

Associated with each issue of both the *Stata Journal* and the STB are the programs and datasets described therein. These programs and datasets are made available for download and installation over the Internet, not only to subscribers, but to all Stata users. See [R] **net** and [R] **sj** for more information.

Because the *Stata Journal* and the STB have had several articles dealing with regression models, if you select **Help > Search...** and choose *Search documentation and FAQs*, and type `ordered regression`, you will see some of these references:

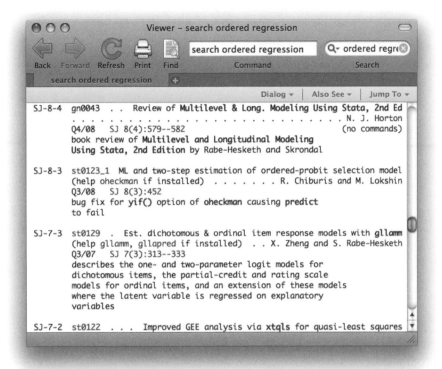

SJ-8-4 refers to volume 8, number 4 of the *Stata Journal*. This is an example of a review article.

SJ-8-3 refers to volume 8, number 3 of the *Stata Journal*. This is an example of a *Stata Journal* article that updates the `oheckman` user-written command. The command can be downloaded and installed to extend the abilities of Stata. See *Downloading user-written programs* in [GSM] **19 Updating and extending Stata—Internet functionality** for more information.

Clicking on the SJ links will open a browser and take you to the *Stata Journal* website, where you can download abstracts and articles. The *Stata Journal* website allows all articles that are older than three years old to be downloaded for free.

Links to other sites where you can freely download programs and datasets for Stata can be found on the Stata website; see http://www.stata.com/links/. See *Downloading user-written programs* in [GSM] **19 Updating and extending Stata—Internet functionality** for more details on how to install this software. Also see [R] **ssc** for information on a convenient interface to resources available from the Statistical Software Components (SSC) archive.

We recommend that all users subscribe to the *Stata Journal*. See [U] **3.5 The Stata Journal** for more information.

Notes

5 Opening and saving Stata datasets

How to load your dataset from disk and save it to disk

Opening and saving datasets in Stata works similarly to those tasks in other computer applications. There are a few differences, however. First off, it is possible to save and open files from within Stata's Command window. Second, Stata allows just one dataset to be open and in use at any one time. It is possible to have many Viewers viewing many files, but only one dataset may be in use at any time. Keeping this in mind will make Stata's care in opening new datasets clear. This chapter outlines all the possible ways to open and save datasets.

A Stata dataset can be opened in a variety of ways, most of which are probably familiar to you from other applications:

- Double-click on a Stata data file, which is a file whose extension is dta. Note: The file extension may not be visible, depending on what options you have set in your operating system.
- Select **File > Open...** or click on the **Open** button and navigate to the file.
- Select **File > Open Recent >** *filename*.
- Type use *filename* in the Command window. Stata will look for *filename* in the current working directory. If the file is located elsewhere, you will need to give its path. Be aware that if there is a space anywhere in the path or filename, you will need to put the filename inside quotation marks. See [U] **11.6 Filenaming conventions**.

Because Stata has at most one dataset open at a time, opening a dataset will cause Stata to discard the dataset that is currently in memory. If there have been changes to the data in the currently open dataset, Stata will refuse to discard the dataset unless you force it to do so. If you open the file with any method other than the Command window, you will be prompted. If you use the Command window and the current data have changed, you will get the following error message:

```
. sysuse auto
no; data in memory would be lost
r(4);
```

These behaviors protect you from mistakenly losing data.

To save an unnamed dataset (or an old dataset under a new name):
1. Select **File > Save As...**.
2. Type save *filename* in the Command window.

To save a dataset for use with Stata 9 or Stata 10 (Stata 11 can load Stata 12 datasets):
1. Select **File > Save As...**, and select **Stata 9/10 Data (*.dta)** from the **Format** pop-up menu.
2. Type saveold *filename* in the Command window.

To save a dataset that has been changed (overwriting the original data file):
1. Select **File > Save**.
2. Click on the **Save** button.
3. Type save, replace in the Command window.

Once you overwrite a dataset, there is no way to recover your original dataset. With important datasets, you may want to either keep a backup copy of your original *filename*.dta or save your changes to a dataset under a new name. This is no different from working with a word-processing document, except that recovering from an inadvertent save of a dataset is nearly impossible.

Important note: Changes you have made to a dataset are not permanent until you save them. You work with a copy of the dataset in memory, not with the data file itself. This should not be surprising, because it is the way that you work with almost all applications on your computer.

If you do not want to save your dataset, you can clear the dataset in memory and open a new dataset by typing use *filename*, clear.

6 Using the Data Editor

The Data Editor

The Data Editor gives a spreadsheet-like view of data, if any, that are currently in memory. You can use it to enter new data, edit existing data, and edit attributes of the data in the dataset, such as variable names, labels, and display formats, as well as value labels.

In addition to the view of the data, there are two windows for manipulating variables and their properties: the Variables window and the Properties window. These are similar to the like-named windows in the main Stata window.

Any action you take in the Data Editor results in a command being issued to Stata as though you had typed it into the Command window. This means that you can keep good records and learn commands by using the Data Editor.

The Data Editor can be kept open while you work in Stata, giving you a live view of your dataset as you work. To protect your data from inadvertent changes, the Data Editor has two modes: edit mode for active editing and browse mode for viewing. In browse mode, editing within the Data Editor window is disabled. We highly recommend that you use the Data Editor in browse mode by default and switch to edit mode only when you want to make changes.

We will be entering and editing data in this chapter, as well as manipulating the variables using the Variables and Properties windows, so start the Data Editor in edit mode by clicking on the **Data Editor (Edit)** button, .

Buttons on the Data Editor

The toolbar for the Data Editor has some standard buttons and some buttons we have not yet seen:

Edit: Changes the Data Editor to edit mode.

Browse: Changes the Data Editor to browse mode for safely looking at data.

Filter: Filters the observations visible in the Data Editor. This button is useful for looking at a subset of the current dataset.

Variables: Toggles the visibility of the Variables window.

Properties: Toggles the visibility of the Properties window.

Snapshots: Opens the Snapshots window. See *Working with snapshots* below.

You can move about in the Data Editor by using the typical methods:

- To move to the right, use the *Tab* key or the right arrow key.
- To move to the left, use *Shift–Tab* or the left arrow key.
- To move down, use *Return* or the down arrow key.
- To move up, use *Shift–Return* or the up arrow key.

You can also click within a cell to select it.

Right-clicking within the Data Editor brings up a contextual menu that allows you to manipulate the data and what you are viewing. We will cover much of this below. Right-clicking on the Data Editor window displays a menu from which you can do many common tasks:

- **Copy** to copy data to the Clipboard.
- **Paste** to paste data from the Clipboard.
- **Select All** to select all the data displayed in the Data Editor. This could be different from the data in the dataset if the data are filtered or some variables are hidden.
- **Add Variable at End of Dataset...** to bring up a dialog for creating a new variable.
- **Replace Contents of Variable...** to bring up a dialog for replacing the values of the selected variable.
- **Sort Data...** to sort the dataset by the selected variable.
- **Value Labels** to access a submenu for managing and displaying value labels.
- **Keep Only Selected Data** to keep only the selected data in the dataset. All remaining data will be dropped (removed) from the dataset. As always, this affects only the data in memory. It will not affect any data on disk.
- **Drop Selected Data** to drop the selected data. This is only possible if the selected data consist of either specific variables or specific observations.
- **Hide Selected Variables** to hide the selected variables.
- **Show Only Selected Variables** to hide all but the selected variables.
- **Show Entire Dataset** to turn off all filters and unhide all variables.
- **Preferences...** to set the preferences for the Data Editor.

Data entry

Entering data into the Data Editor is similar to entering data into a spreadsheet. One major difference is that the Data Editor has the concept of observations, which makes the data entry smart. We will illustrate this with an example. It will be useful for you to follow the example at your computer. To work along, you will need to start with an empty dataset to work along, so save your dataset if necessary, and then type `clear` in the Command window.

Note: As a check to see if your data have changed, type `describe, short` (or `d,s` for short). Stata will tell you if your data have changed.

Suppose that we have the following data, and we want to enter it into Stata:

Make	Price	MPG	Weight	Gear Ratio
VW Rabbit	4697	25	1930	3.78
Olds 98	8814	21	4060	2.41
Chev. Monza	3667		2750	2.73
AMC Concord	4099	22	2930	3.58
Datsun 510	5079	24	2280	3.54
	5189	20	3280	2.93
Datsun 810	8129	21	2750	3.55

We do not know MPG for the third car or the make of the sixth.

Start by opening the Data Editor in edit mode. You can do this either by clicking on the **Data Editor (Edit)** button, 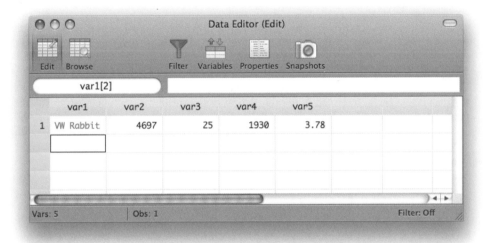, or by typing `edit` in the Command window. You should be greeted by a Data Editor with no data displayed. (If you see data, type `clear` in the Command window.) Stata shows the active cell by highlighting it and displaying *varname*[*obsnum*] next to the input box in the Cursor Location box. We will see below that we can navigate within a dataset by using this cell reference. The Data Editor starts, by default, in the first row of the first column. Because there are no data, there are no variable names, and so Stata shows `var1[1]` as the active cell.

We can enter these data either by working across the rows (observation by observation) or by working down the columns (variable by variable). To enter the data observation by observation, press *Tab* after entering each value until you have reached the end of the first row. In our case, we would type `VW Rabbit`, press *Tab*, type `4697`, press *Tab*, and continue entering data to complete the first observation.

After you are finished with the first observation, select the second cell in the first column, either by clicking within it or by navigating to it. At this point, your screen should look like this:

We can now enter the data for the second observation in the same fashion as the first—with one nice difference: after entering the last value in the row, pressing the *Tab* key will bring us to the first cell in the third row. This is possible because the number of variables is known after the first observation has been entered, so Stata knows when it has all the data for an observation.

We can enter the rest of the data by pressing the *Tab* key between entries, simply skipping over missing values by tabbing through them.

If we had wanted to enter the data variable by variable, we could have done that by pressing *Return* between each make of car until all seven observations were entered, skipping past the missing entry by pressing *Return* twice in succession. Once the first variable was entered, we would select the first cell in the second column and enter the price data. We would continue this until we were finished.

Notes on data entry

There are several things to note about data entry and the feedback you get from the Data Editor as you enter data:

- Stata does not allow blank columns or rows in the middle of your dataset.
  Whenever you enter new variables or observations, always begin in the first empty column or row. If you skip columns or rows, Stata will fill in the intervening columns or rows with missing values.

- Strings and value labels are color coded.
  To help distinguish between the different types of variables in the Data Editor, string values are displayed in red, value labels (see [GSM] **9 Labeling data**) are displayed in blue, and all other values are displayed in black. You can change the colors for strings and value labels by right-clicking on the Data Editor window and selecting **Preferences...**.

- A period (.) represents Stata's system missing numeric value.
  Stata has a system missing value, '.', and extended missing values '.a' through '.z'. By default, Stata uses its system missing value.

- The *Tab* key is smart.
  As we saw above, after the first observation has been entered, Stata knows how many variables you have. So at the end of the second observation (and all subsequent observations), *Tab* will automatically take you back to the first column.

- The Cursor Location box both shows location and is used for navigation.
  The Cursor Location box gives the location of the current cell. If you see, for example, var3[4], this means that the current cell is the fourth observation of the variable named var3. You can navigate to a particular cell by typing the variable name and the observation in the Cursor Location box. If you wanted the second observation of var1 to be the active cell, typing var1 2 in the Cursor Location box and pressing *Return* would take you there.

- Double quotes around text are unnecessary in string variables.
  Once Stata knows that a variable is a string variable (it holds text), there is no need to put quotes around the values, even if the values look like a number. Thus, if you wanted to enter ZIP codes as text, you would enter the first ZIP code with quotes ("02173"), but the rest would not need any quotes.

- If you select a cell and type new data, using an arrow key will accept the change and move to a new active cell. If you double-click on a cell, you can edit within the cell contents. In this case, the arrow keys move within the cell's data.

- If, while you are entering data in a cell, you decide you would like to cancel the changes, press the *Esc* key or click outside the cell.

Renaming and formatting variables

The data have now been entered into Stata, but the variable names leave something to be desired: they have the default names var1, var2, ..., var5. We would like to rename the variables so that they match the column titles from our dataset. We would also like to give the variables descriptions and change their formatting.

We will step through changing the name, label, and format of the price variable. We will then add a note to the variable. Start by clicking on the var2 variable in the Variables window. The few properties associated with var2 are now visible and editable in the Properties window. We may now systematically change the properties of var2 to our choosing:

1. Double-click on var2 in the *Name* field to select the old variable name, and type price to overwrite the name.

2. Double-click under the new `price` name in the *Label* field.

3. Enter a worthwhile label, such as `Price in Dollars`.

4. Click on `%9.0g` in the *Format* field.

5. Click on the ellipses (...) button that appears. The *Create Format* dialog opens.

6. You can see here that there are many possible formats, most of which are related to time. We want commas in our numbers, so check the *Use commas in numeric output* checkbox. When you are done, click on the **OK** button.

7. Click in the *Notes* field.

8. Click on the ellipses (...) button that appears. A dialog called *Notes for price* opens.

9. Click on the **Add** button and type a clever note.

10. When you are done typing, click on the **Submit** button, and then click the on **Close** button. This note is now attached to the `price` variable.

11. Click on the expose triangle to see the note you just typed in the Properties window.

To edit the properties of another variable, either click on the variable in the Variables window or use the navigation arrows in the *Properties* window until the variable's name appears. We can name the first variable `make`; the third, `mpg`; the fourth, `weight`; and the fifth, `gear_ratio`. Just before you rename `var5` to `gear_ratio`, your screen should look like this:

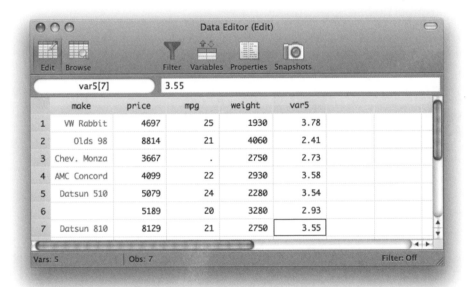

You need to know some rules for variable names:

• Stata is case sensitive.
  `Make`, `make`, and `MAKE` are all different names to Stata. If you had named your variables `Make`, `Price`, `MPG`, etc., then you would have to type them correctly capitalized in the future. Using all lowercase letters is easier.

• A variable name must be 1–32 characters long.

• The characters can be letters (A–Z, a–z), digits (0–9), or underscores (_).

• Spaces or other characters are not allowed.

• The first character of a variable name must be a letter or an underscore.
  Although you can use an underscore to begin a variable name, it is highly discouraged. Such

names are used for temporary variable names in Stata. You would like your data to be permanent, so using a temporary name could lead to great frustration.

For more information about variable names and value labels, see [GSM] **9 Labeling data**; for display formats, see [U] **12.5 Formats: Controlling how data are displayed**.

Copying and pasting data

You can copy and paste data by using the Data Editor. This is often a simple way to bring data into Stata from any other applications such as spreadsheets or databases.

1. Select the data that you wish to copy using one of these means:
   - Click once on a variable name or column heading to select an entire column.
   - Click once on an observation number or row heading to select the entire row.
   - Click and drag the mouse to select a range of cells.
2. Copy the data to the Clipboard by right-clicking within the selected range, and select **Copy**.
3. Paste the data from the Clipboard by clicking on the top left cell of the area to which you wish to paste. Then right-click on the same cell, and select **Paste**.

We will illustrate copying and pasting an observation by making a copy of the first observation and pasting it at the end of the dataset.

Start by clicking on the observation number of the first observation. Doing so highlights all the data in the row. Right-click on the same location (there is no need to move the mouse), and select **Copy**:

Click on the first cell in the eighth row, right-click while you are still in that cell, and choose **Paste** from the resulting menu. You can see that the observation was successfully duplicated.

Notes on copying and pasting

- The above example illustrated copying and pasting within the Data Editor. You can use roughly the same technique to copy and paste between other applications and Stata, and between Stata and other applications. The easiest way to see if copying and pasting works properly with another application is to try it. The one requirement for things to work well is that the external application must copy tables in either tab-delimited or comma-delimited form, as do spreadsheet applications, many database applications, and some word processors. Using **Edit > Paste Special...** gives some added flexibility to the formats you can paste into the Data Editor. If a simple paste does not give you what you expected, you should try **Edit > Paste Special...**. For more information on file-based methods for importing data into Stata, see [GSM] **8 Importing data**.

- If you are copying and pasting data with value labels, you have a choice. You can copy variables with value labels as text, using the value labels as the actual values, or you can copy said variables as their underlying encoded numbers. Copying with the value labels is the default. If you would like the other choice, right-click in the Data Editor and select **Value Labels > Hide All Value Labels**.

Changing data

As its name suggests, the Data Editor can be used to edit your dataset. As we have seen already, it can be used to edit the data themselves as well as the description and display options for the variables.

Here is an example for making some changes to the `auto` dataset, which illustrates both methods for using the Data Editor and its documentation trail. We will also keep snapshots of the dataset as we are working, so that we can revert to previous versions of the dataset in case we make a mistake.

We would like to investigate the dataset, work with value labels, delete the `trunk` variable, and make a new variable showing gas consumption per 100 miles. These tasks will illustrate the basics of working in the Data Editor.

Start by typing `sysuse auto` into the Command window. You get an error and are told that your data have changed. This is good—Stata is keeping you from inadvertently throwing away the unsaved changes to your current data as it loads the `auto` dataset. If you would like to save the dataset you have been working on, select **File > Save** and save the dataset in an appropriate location. Otherwise, type `clear` in the Command window and press *Return* to clear out the data.

Once the `auto` dataset is loaded, start the Data Editor.

1. We would like to see which cars have the lowest and highest gas mileages. To do this, right-click within the `mpg` column. Select **Sort Data...** from the contextual menu. A dialog will pop up asking if you want to sort. Click on **OK**. (Stata worries about sort order, because sort order can affect reproducibility when using resampling techniques. This is a good thing.) You will see that the data have now been sorted by `mpg` in ascending order. The lowest-mileage cars are at the top of the screen; by scrolling to the bottom of the dataset, you can find the highest-mileage cars. For more general sorting, you could use **Data > Sort**, and pick what you need.

2. We would like to investigate repair records, and hence sort by the rep78 variable. (Do this now.) We see that the Starfire and Firebird both had poor repair records, but we would like to see the cars with good repair records. We could scroll to the bottom of the dataset, but it will be faster to use the Cursor Location box: type rep78 74 and press *Return* to make rep78[74] the active cell. We notice that the last five entries for rep78 appear as dots. The dots mean that these values are missing. A few items of note:

   - As we can see from the result of the sort, Stata views missing values as being larger than all numeric nonmissing values. In technical terms, this means that rep78 >= . is equivalent to missing(rep78).
   - What we do not see here is that Stata has multiple missing-value indicators: . is Stata's default or system missing-value indicator, and .a, .b, ..., .z are Stata's extended missing values. Extended missing values are useful for indicating the reason why a value is unknown.
   - The different missing values sort among themselves: . < .a < .b < ⋯ < .z. See [U] **12.2.1 Missing values** for full details.

3. We would like to make the repair records readable. Click on rep78 in the Variables window.
4. Click on the *Value Label* field in the Properties window, and then click on the ellipses (...) button that appears. This opens the *Manage Value Labels* dialog. We need to define a new value label for the repair records.

   a. Click on the **Create Label** button. You will see the *Create Label* sheet.
   b. Type a name for the label, say, repairs, in the *Label name* box.
   c. Press the *Tab* key or click within the *Value* field.
   d. Type 1 for the value, press the *Tab* key, and type atrocious for the label.
   e. Press the *Return* key to create the pairing.
   f. Repeat steps d and e to make all the pairings: 2 with "bad", 3 with "OK", 4 with "good", and 5 with "stupendous".
   g. Click on the **OK** button to finish creating the value label.
   h. Click on the expose button, ▶, to show the label—you should see this:

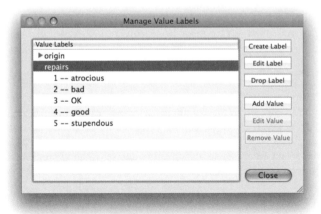

   If you have something else, you can edit the label by clicking on the **Edit Label** button.
   i. Click on the **Close** button to close the *Manage Value Labels* dialog.

Now that the label has been created, attach it to the rep78 variable by clicking on the double arrows in the *Value Label* field and selecting the **repairs** menu item. You can now see the labels displayed in place of the values.

5. Suppose that we found the original source of the data in a time capsule, so we could replace some of the missing values for `rep78`. We could type the values into cells. We can also assign the values by right-clicking within a cell with a missing value and choosing a value from **Value Labels > Assign Value from Value Label 'repairs'**. This strategy can be useful when a value label has many possible values.

6. We would now like to delete the `trunk` variable. We can do this by right-clicking on the `trunk` variable name at the top of the column and selecting the **Drop Selected Data** menu item. Because this can lead to data loss, the Data Editor asks whether we would like to drop the selected variable. Click on the **Yes** button.

7. To finish up, we would like to create a variable containing the gallons of gasoline per 100 miles driven for each of the cars.

   a. Right-click within any cell and choose the **Add Variable at End of Dataset...** menu item to bring up the *generate* dialog.

   b. Type `gp100m` in the *Variable name* field.

   c. Being sure that the *Specify a value or an expression* radio button is selected, type `100/mpg` in its field. We could have clicked on the **Create...** button to open the *Expression builder* dialog, but this formula was simple enough to type. (You might want to explore the Expression builder right now to see what it can do.)

   d. Click on **OK**. You can scroll to the right to see the newly created variable.

Throughout this data editing session, we have been using the Data Editor to manipulate the data. If you look in the Results window, you will see the commands and their output. You can also see all the commands generated by the Data Editor in the Review window. If you wanted to save the editing commands to use again later, you could do the following steps:

1. Click in the Review window on the last command that came from the Data Editor.

2. Scroll up until you find the `sort mpg` command you ran immediately after opening the Data Editor, and *Shift*-click on it.

3. Right-click on one of the highlighted commands.

4. Select **Send to Do-file Editor**.

This procedure will save all the commands you highlighted into the Do-file Editor. You could then save them as a do-file, which you could run again later. We will talk more about the Do-file Editor in [GSM] **13 Using the Do-file Editor—automating Stata**. You can find help about do-files in [U] **16 Do-files**.

If you want to save this dataset, save it under a new name by using **File > Save As...** to prevent overwriting the original dataset.

Working with snapshots

The Data Editor allows you to save to disk snapshots of whatever dataset you are working on. These are temporary copies of the dataset—they will be deleted when you quit Stata, so they need to be treated as temporary. Still, there are many uses for snapshots, such as

- saving a temporary copy of the data in memory so that another dataset can be opened and viewed;
- saving stages of work, which can be recovered in case you do something disastrous; and
- saving pieces of datasets while doing analyses.

We will keep using the `auto` dataset from above; if you are starting here, you can start fresh by typing `sysuse auto` in the Command window to open the dataset. (If you get a warning about data in memory being lost, either use `clear` or save your data. See [GSM] **5 Opening and saving Stata**

datasets for more information.) If we open the Data Editor and click on the **Snapshots** button, 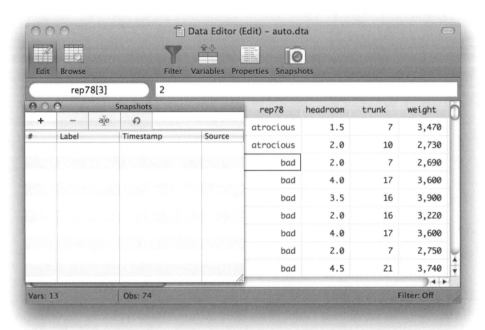, we see the following window. If you are starting afresh, you will see numbers rather than labels for rep78.

To begin with, only one button is active. Click on the active button—the **Add** button, ⊞ . It brings up a dialog asking for a label, or name, for the snapshot. Give it an inventive name, such as Start. You can see that a snapshot is now listed in the Snapshots window, and all the buttons are highlighted. The following buttons appear in the Snapshots window.

+ **Add:** Save a new snapshot with a timestamp and label.

− **Remove:** Erase a snapshot. This action deletes the temporary snapshot file, but does not affect the data in memory.

aȷe **Change Label:** Edit the label of the selected (highlighted) snapshot.

↻ **Restore:** Replace the data in memory with the data from the selected snapshot. If the data in memory have changed, you will get a dialog confirming your action.

You should now try manipulating the dataset using the tools we have seen. Once you have done that, create another snapshot, calling it Changed. Open the Snapshots window and restore the Start snapshot by clicking first on it and then on the **Restore** button to see where you started. You can then go back to where you were working by restoring your Changed snapshot.

Snapshots continue to be available either until they are deleted or you quit Stata. You can thus use snapshots of one dataset while working on another. You will find your own uses for snapshots—just take care to save datasets you want for future use.

Dates and the Data Editor

The Data Editor has two special tools for working with dates in Stata. To see these in action, we will need to open another dataset. Either save your dataset or `clear` it out, and then type `sysuse sp500` in the Command window. Look in the Data Editor to see what you have.

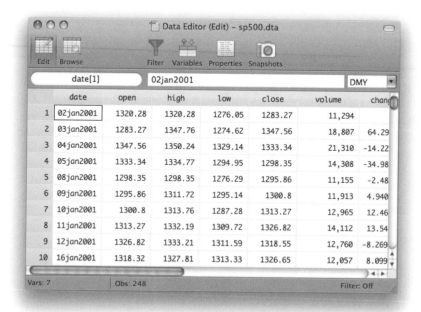

You can see that there is a `date` variable that has January 2, 2001, as its first day, but it is in Stata's default format for dates. You should also notice in the upper-right corner, the *date mask* field, which shows DMY. This field affects how dates are entered when editing data.

We will start with formatting:

1. Select the `date` variable in the Variables window to the right of the data table.
2. In the Properties window, select the Format row and click on the ellipses button that appears.
3. The *Create Format* dialog tells us three pieces of information about the date format:
   - These are daily dates. As you can see, Stata understands other types of dates that are often used in financial data.
   - Looking at the bottom of the dialog, you can see that the Stata default date format is `%td`. This means that the variable contains time values that are to be interpreted as daily dates.
   - This default format is displayed as, for example, `07apr2009`. (Of course, this was clear in the Data Editor.)
4. There are many premade date formats in the *Samples* pane at the top right of the *Create Format* dialog. Click on `April 07, 2009`. You can see how the format would be specified at the bottom of the dialog.
5. Click on **OK** to close the *Create Format* dialog. You can see that the dates are now displayed differently.

This is a very simple way to change date formats. For complete information on dates and date formats, see [D] **datetime**.

We will now change some of the dates to illustrate how the date mask works. By default, the date mask is set to DMY. This means dates can be entered in many different fashions, as long as the order is day, month, year. Try the following:

1. Click in the first observation of date, so that the Cursor Location shows date[1]
2. Type 18jan2009 and press the *Return* key. Stata understands the DMY date mask and knows enough to enter the new date in the selected cell.
3. Enter 30042010 and press *Return*. Stata still understands the date mask, even though there are no separators.
4. Click within the *date mask* field, and choose MDY from the drop-down menu.
5. Click on any observation in the date column.
6. Type March 15, 2011 and press *Return*. Stata will still understand.

Working in this fashion is the fastest way to edit dates by hand. If you look in the Results window, you will see why.

We are now finished with this dataset, so type clear and press *Return*.

Data Editor advice

As you could see above, a small mistake in the Data Editor could cause large problems in your dataset. You really must take care in how you edit your data.

- People who care about data integrity know that editors are dangerous—it is easy to accidentally make changes. **Never use the Data Editor in edit mode when you just want to look at your data.** Use the Data Editor in browse mode (or use the browse command).

- If you must edit your data, protect yourself by limiting the dataset's exposure. For example, if you need to change rep78 only if it is missing, find a way to look at just the missing values for rep78 and any other variables needed to make the change. This will make it impossible for you to change (damage) variables or observations other than those you view. We will explore this aspect shortly.

- Even with these caveats, Stata's Data Editor is safer than most because it records commands in the Results window. Use this feature to log your output and make a permanent record of the changes. Then you can verify that the changes you made are the changes you wanted to make. See [GSM] **16 Saving and printing results by using logs** for information on creating log files.

Filtering and hiding

We would now like to investigate restricting our view of the data we see in the editor. This feature is useful for the reasons mentioned above, and as we will see, it helps if we would like to browse through the data of a large dataset. In any case, we would like to focus on some, not all the data, whether we focus on some of the variables, some of the observations, or even just some observations within some variables. We would also like to change the order of the variables. We will show you how this is done by using both the interface and commands.

Open the auto dataset by typing sysuse auto. If you get an error message, type clear and try again. Once you have done that, open the Data Editor.

Suppose that we would like to edit only those observations for which `rep78` is missing. We will need to look at the make of the car so that we know which observations we are working with, but we do not need to see any other variables. We will work as though we had a very large dataset to work with.

1. Before we get started, try experimenting with the Variables window.

    a. Drag variables up and down the list. Doing so changes the order that the variables' columns are displayed in the Data Editor. It does not change their order in the dataset itself.

    b. Uncheck some of the checkboxes in the first column to hide some of the variables.

    c. Type a search criterion in the search field. Just like in the Variables window in the main Stata window, the default is to ignore case and find any variables containing any of the words in the filter. Clicking on the arrow by the magnifying glass will allow you to change this behavior. The filtering of variables in the list affects what is displayed in the Variables window; it does not affect what variables' data are displayed. When you are done, delete your filter text.

2. Right-click on any variable in the Variables window and select **Select All** from the contextual menu.

3. Click on any checkbox to deselect all the variables.

4. Click on the `make` variable to select it and deselect all the other variables.

5. Click on the checkbox for `make`.

6. Click on the checkbox for `rep78`.

If you look in the Command window, you can see that no commands have been issued, because hiding the variables does not affect the dataset—it affects only what shows in the Data Editor.

We now have protected ourselves by using only those variables that we need. We should now reduce our view to only those observations for which `rep78` is missing. This is simple.

1. Click on the **Filter Observations** button, , in the Data Editor's toolbar.

2. Enter `missing(rep78)` in the *Filter by expression* field.

3. Click on the **Apply Filter** button.

4. If you are curious, click the ellipses button. It opens up an *Expression builder* dialog. This lists the wide variety of functions available in Stata. See [D] **functions**.

Now we are focused on the part of the dataset in which we would like to work, and we cannot destroy or mistakenly alter other data by stray keystrokes in the Data Editor window.

It is worth learning how to hide variables and filter observations in the Data Editor from the Command window. This can be quite convenient if you are going to restrict your view, as we did above. To work from the Command window, we must use the `edit` command together with a varlist (variable list) along with `if` and `in` qualifiers in the Command window. By using a varlist, we restrict the variables we look at, whereas the `if` and `in` qualifiers restrict the observations we see. ([GSM] **10 Listing data and basic command syntax** contains many examples of using a command with a variable list and `if` and `in`.) Suppose we want to correct the missing values for `rep78`. The minimum amount of data we need to expose are `make` and `rep78`. To see this minimal amount of information and hence to minimize our exposure to making mistakes, we enter the commands

```
. sysuse auto
(1978 Automobile Data)
. edit make rep78 if missing(rep78)
```

and we would see following window:

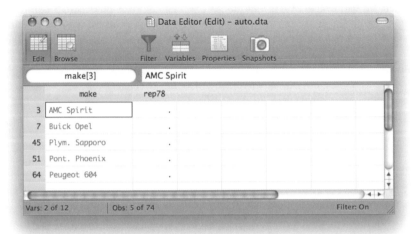

Once again, we are safe and sound.

Keep this lesson in mind if you edit your data. It is a lesson well learned.

Browse mode

The purpose of using the Data Editor in browse mode is to be able to look at data without altering it by stray keystrokes. You can start the Data Editor in browse mode by clicking on the **Data Editor (Browse)** button, [icon], or by typing browse in the Command window. When working in browse mode, all contextual menu items that would let you alter the data, the labels, or any of the display formats for the variables are disabled. You may view a variable's properties with the **Variable Properties...** menu item, but you may not make any changes. You still can filter observations and hide variables to get a restricted view, because these actions do not change the dataset.

Note: Because you can still use Stata menus not related to the Data Editor and because you can still type commands in the Commands window, it is possible to change the data even if the Data Editor is in browse mode. In fact, this means you can watch how your commands affect the dataset. You are merely restricted from using the Data Editor itself to change the data.

7 Using the Variables Manager

The Variables Manager

This chapter discusses Stata's Variables Manager.

The Variables Manager is a tool for managing properties of variables both individually and in groups. It can be used to create variable and value labels, rename variables, change display formats, and manage notes. It has the ability to filter and group variables as well as to create variable lists. Users will find these features useful for managing large datasets.

Any action you take in the Variables Manager results in a command being issued to Stata as though you had typed it in the Command window. This means that you can keep good records and learn commands by using the Variables Manager.

The Variable pane

The Variable pane shows the list of variables in the dataset. This list can be manipulated in a variety of ways.

- The variables can be filtered by entering text into the filter box in the upper-left corner. This can be a good way to zoom in on similarly named or labeled variables.
- The list can be sorted by clicking on the column title.
    a. If you click on a column title, it will sort in ascending order.
    b. A second click on the same column title will change to sorting in descending order.
    c. A third click on the same column will restore the original dataset order.

The sort order affects only the how the data appear in Variable Manager window—the dataset itself stays the same.

• The order of the columns can be changed by dragging the column titles. To restore the original column headings, right-click on the column titles and select **Restore Column Defaults**.

Right-clicking on the Variable pane

Right-clicking on the Variable pane displays a menu from which you can do many common tasks:
• **Keep Only Selected Variables** to keep only the selected variables in the dataset and to drop all the others.
• **Drop Selected Variables** to drop all the selected variables from the dataset.
• **Manage Notes for Selected Variable...** to open a window that allows adding and deleting notes for a single variable. This is disabled if multiple variables are selected.
• **Manage Notes for Dataset...** to open a window that allows adding and deleting notes for the dataset as a whole.
• **Copy Varlist** to copy the names of the selected variables to the Clipboard.
• **Select All** to select all visible variables. If a variable has become hidden because of the filter, it will not be selected.
• **Send Varlist to Command Window** to insert the names of the selected variables in the Command window. Combined with grouping and sorting, this can be a useful way to create variable lists in large datasets.

The Variable Properties pane

The Variable Properties pane can be used to manipulate the properties of variables selected in the Variable pane. With one variable selected, you can manipulate all properties of the variable. With many variables selected, you can change their formats or types, as well as assign value labels all at once. These fields work in the same fashion as those shown in *Renaming and formatting variables* in [GSM] **6 Using the Data Editor**. We can also manage the notes Stata allows you to attach to variables and the dataset—we will show an example below.

Managing notes

Stata allows you to attach notes to both variables and the dataset as a whole. These are simple text notes that you can use to document whatever you like—the source of the dataset, data collection quirks associated with a variable, what you need to investigate about a variable, or anything else.

Start by selecting a variable in the Variable pane. We will work with the `price` variable. Click on the **Manage...** button next to the *Notes* field, and you will see the following dialog appear:

We will add a few notes:

1. Click on the **Add** button to add a note.
2. Type `TS - started working`. TS with a trailing space inserts a timestamp in the note.
3. Add two more notes. We added two notes about prices:

It is worth experimenting with adding, deleting, and editing notes. Notes can be an invaluable memory aid when working on projects that last over a long period of time. Anytime you manipulate notes in the Notes Manager, you create Stata commands.

Notes

8 Importing data

Copying and pasting

One of the easiest ways to get data into Stata is often overlooked: you can copy data from most applications that understand the concept of a table and then paste the data into the Data Editor. This approach works for all spreadsheet applications, many database applications, some word-processing applications, and even some webpages. Just copy the full range of data, paste it into the Data Editor, and everything will probably work well. You can even copy a text file that has the pieces of data separated by commas and then paste it into the Data Editor.

Suppose that your friend has a small dataset about some very old cars.

```
VW Rabbit,4697,25,1930,3.78
Olds 98,8814,21,4060,2.41
Chev. Monza,3667,,2750,2.73
,4099,22,2930,3.58
Datsun 510,5079,24,2280,3.54
Buick Regal,5189,20,3280,2.93
Datsun 810,8129,,2750,3.55
```

You would like to put these data into Stata. Doing so is easier than you think:

1. Clear out your current dataset by typing `clear`.
2. Copy the data from the PDF documentation the way you would copy anything from any document. (For best results use Adobe Reader.)
3. Open the Data Editor in edit mode.
4. Select **Edit > Paste Special**.
5. Stata sees that the column delimiters are commas and shows how the data would look.
6. Click on the **OK** button.

You can see that Stata has imported the data nicely.

Later in this chapter, we would like bring these data into Stata without copying and pasting, so we would like to save them as a text file. Go back to the main Stata window and click on the **Do-file Editor** button ![Do-file Editor] to open a new Do-file Editor window. Paste the data in the Do-file Editor, then click on the **Save** button. Navigate to your working directory and save the file as `a few cars.csv`. If you do not know what your working directory is, look in the status bar at the bottom of the main Stata window.

Be careful if you are copying data from a spreadsheet because spreadsheets can contain special formatting that ruins its rectangular form. Be sure that your spreadsheet does not contain blank rows, blank columns, repeated headers, or merged cells because these can cause trouble. As long as your spreadsheet looks like a table, you will be fine.

Commands for importing data

Copying and pasting is a great way to bring data into Stata, but if you need a clear audit trail for your data, you will need another way to bring data into Stata. The rest of this chapter will explain how to do this. You will also learn methods that lend themselves better to repetitive tasks and methods for importing data from a wide variety of sources.

Stata has various commands for importing data. The three main commands for reading non–Stata datasets in plain text (ASCII) are

- insheet, which is made for reading text files created by spreadsheet or database programs;
- infile, which is made for reading simple data that are separated by spaces or rigidly formatted data aligned in columns; and
- infix, which is made for data aligned in columns, but possibly split across rows.

Stata has other commands that can read other types of files and can even get data from external databases without the need for an interim file:

- The import excel command can read Microsoft Excel files directly, either as .xls or as .xlsx files.
- The import sasxport command can read any SAS XPORT file, so data can be transferred from SAS to Stata in this fashion.
- The odbc command can be used to pull data directly from any data sources for which you have ODBC drivers.
- The xmluse command can read some XML files (most notably, Microsoft Excel's Spread-sheetML).

Each command expects the file that it is reading to be in a specific format. This chapter will explain some of those formats and give some examples. For the full story, consult the *Data-Management Reference Manual*.

The insheet command

The insheet command was specially developed to read in text (ASCII) files that were created by spreadsheet or database programs because these are common formats for sharing datasets on the Internet. All spreadsheet programs and most database applications have an option to save the dataset as a text (ASCII) file with the columns delimited with either tab characters or commas. Some of these programs also save the column titles (variable names, in Stata) in the text file.

To read in such a file, you have only to type insheet using *filename*, where *filename* is the name of the text file. The insheet command will figure out what the delimiter character is (tab or comma) and what type of data is in each column. As always, if *filename* contains spaces, put double quotes around the filename, and include the path if *filename* is not in the current working directory.

By default, the insheet command understands files that use the tab or comma as the column delimiter. If you have a file that uses a space as the delimiter, use the infile command instead; see [D] **infile (free format)** for more information. If you have a file that uses another character as the delimiter, use insheet's delimiter() option; see [D] **insheet** for more information.

Earlier in this chapter, you saved a file called a few cars.csv whose contents were shown as the small amount of data in *Copying and pasting*. These data correspond to the make, price, MPG, weight, and gear ratio of a few very old cars. The variable names are not in the file (so insheet will assign its own names), and the fields are separated by commas. Clear out any existing data, then use insheet to read the data in this file. Because there are spaces in the filename, it must be enclosed in double quotes.

```
. clear
. insheet using "a few cars.csv"
(5 vars, 7 obs)
```

You can look at the data in the Data Editor, and it will look just like the earlier result from copying and pasting. We will now list the data so that we can see them in the manual. The `separator(0)` option suppresses the horizontal separator line that is drawn after every fifth observation by default.

```
. list, separator(0)

             v1      v2   v3     v4     v5

  1.    VW Rabbit    4697   25   1930   3.78
  2.      Olds 98    8814   21   4060   2.41
  3.   Chev. Monza   3667    .   2750   2.73
  4.                 4099   22   2930   3.58
  5.    Datsun 510   5079   24   2280   3.54
  6.   Buick Regal   5189   20   3280   2.93
  7.   Datsun 810    8129    .   2750   3.55
```

If you want to specify better variable names, you can include the desired names in the command:

```
. insheet make price mpg weight gear_ratio using "a few cars.csv"
(5 vars, 7 obs)
. list, separator(0)

             make   price   mpg   weight   gear_r~o

  1.    VW Rabbit    4697    25    1930      3.78
  2.      Olds 98    8814    21    4060      2.41
  3.   Chev. Monza   3667     .    2750      2.73
  4.                 4099    22    2930      3.58
  5.    Datsun 510   5079    24    2280      3.54
  6.   Buick Regal   5189    20    3280      2.93
  7.   Datsun 810    8129     .    2750      3.55
```

As a side note about displaying data, Stata listed `gear_ratio` as `gear_r~o` in the output from list. `gear_r~o` is a unique abbreviation for the variable `gear_ratio`. Stata displays the abbreviated variable name when variable names are longer than eight characters.

To prevent Stata from abbreviating `gear_ratio`, you could specify the `abbreviate(10)` option:

```
. list, separator(0) abbreviate(10)

             make   price   mpg   weight   gear_ratio

  1.    VW Rabbit    4697    25    1930        3.78
  2.      Olds 98    8814    21    4060        2.41
  3.   Chev. Monza   3667     .    2750        2.73
  4.                 4099    22    2930        3.58
  5.    Datsun 510   5079    24    2280        3.54
  6.   Buick Regal   5189    20    3280        2.93
  7.   Datsun 810    8129     .    2750        3.55
```

For more information on the ~ abbreviation and on list, see [GSM] **10 Listing data and basic command syntax**.

We will use this dataset again in the next chapter, so we would like to save it. Type save afewcars and press *Return* in the Command window to save the dataset.

For this simple example, you could have copied the above file and pasted it into the Data Editor using **Paste > Special...** and choosing comma as the delimiter.

For text files that have no nice delimiters or for which observations could be spread out across many lines, Stata has two more commands: infile and infix. See [D] **import** for more information about how to read in such files.

Importing files from other software

Stata has some more specialized methods for reading data that were created by other applications and stored in their proprietary formats.

The import excel command is made for reading files created by Microsoft Excel. See import excel in [D] **import excel** for full details.

The import sasxport command can read and create SAS XPORT Transport files. See [D] **import sasxport** for full details.

If you have software that supports ODBC (*Open Database Connectivity*), you can read data by using the odbc command without the need to create interim files. See [D] **odbc** for full details. The FAQ for setting up ODBC is also helpful; read it at http://www.stata.com/support/faqs/data/odbcmu.html.

If you would like to move data by using XML (*Extensible Markup Language*), Stata has the xmluse and xmlsave commands available. See xmluse and xmlsave in [D] **xmlsave** for full details.

Here is a brief summary of the choices:
- If you have a table, you could try copying it and pasting into the Data Editor.
- If you have a Microsoft Excel .xls or .xlsx file, use import excel.
- If you have a file exported from a spreadsheet or database application to a tab-delimited or CSV file, use insheet.
- If you have a free-format, space-delimited file, use infile with variable names.
- If you have a fixed-format file, either use infile with a dictionary or use infix.
- If you have a SAS XPORT file, use import sasxport.
- If you have a database accessible with ODBC, use odbc.
- If you have an XML file, use xmluse.
- Finally, you can purchase a transfer program that will convert the other software's data file format to Stata's data file format. See [U] **21.4 Transfer programs**.

9 Labeling data

Making data readable

This chapter discusses, in brief, labeling of the dataset, variables, and values. Such labeling is critical to careful use of data. Labeling variables with descriptive names clarifies their meanings. Labeling values within numerical categorical variables ensures that the real-world meanings of the encodings are not forgotten. These points are crucial when sharing data with others, including your future self. Labels are also used in the output of most Stata commands, so proper labeling of the dataset will make much more readable results. We will work through an example of properly labeling a dataset, its variables, and the values of one encoded variable.

The dataset structure: The describe command

At the end of *The insheet command* in [GSM] **8 Importing data**, we saved a dataset called afewcars.dta. We will put this dataset into a shape that a colleague would understand. Let's see what it contains.

```
. use afewcars

. list, separator(0)

              make   price   mpg   weight   gear_r~o

  1.     VW Rabbit    4697    25     1930       3.78
  2.        Olds 98    8814    21     4060       2.41
  3.    Chev. Monza    3667     .     2750       2.73
  4.                   4099    22     2930       3.58
  5.     Datsun 510    5079    24     2280       3.54
  6.    Buick Regal    5189    20     3280       2.93
  7.     Datsun 810    8129     .     2750       3.55
```

The data allow us to make some guesses at the values in the dataset, but, for example, we do not know the units in which the price or weight is measured, and the term "mpg" could be confusing for people outside the United States. Perhaps we can learn something from the description of the dataset. Stata has the aptly named describe command for this purpose (as we saw in [GSM] **1 Introducing Stata—sample session**).

```
. describe
Contains data from afewcars.dta
  obs:            7
  vars:           5                           25 Apr 2011 13:50
  size:         238

              storage  display    value
variable name  type    format     label     variable label

make           str18   %18s
price          float   %9.0g
mpg            float   %9.0g
weight         float   %9.0g
gear_ratio     float   %9.0g

Sorted by:
```

Though there is precious little information that could help us as a researcher, we can glean some information here about how Stata thinks of the data from the first three columns of the output.

1. The *variable name* is the name we use to tell Stata about a variable.

2. The *storage type* (otherwise known as the *data type*) is the way in which Stata stores the data in a variable. There are six different storage types, each having its own memory requirement:

   a. For integers:

      byte for integers between -127 and 100 (using 1 byte of memory per observation)

      int for integers between $-32,767$ and $32,740$ (using 2 bytes of memory per observation)

      long for integers between $-2,147,483,647$ and $2,147,483,620$ (using 4 bytes of memory per observation)

   b. For real numbers:

      float for real numbers with 8.5 digits of precision (using 4 bytes of memory per observation)

      double for real numbers with 16.5 digits of precision (using 8 bytes of memory per observation)

   c. For strings (text) between 1 and 244 characters (using 1 byte of memory per observation per character):

      str1 for one-character-long strings

      str2 for two-character-long strings

      str3 for three-character-long strings

      . . .

      str244 for 244-character-long strings

   Storage types affect both the precision of computations and the size of datasets. A quick guide to storage types is available at help datatypes or in [D] **data types**.

3. The *display format* controls how the variable is displayed; see [U] **12.5 Formats: Controlling how data are displayed**. By default, Stata sets it to something reasonable given the storage type. We would like to make this dataset into something containing all the information we need.

To see what a well-labeled dataset looks like, we can take a look at a dataset stored at the Stata Press repository. We need not load the data (and disturb what we are doing); we do not even need a copy of the dataset on our machine. (You will learn more about Stata's Internet capabilities in [GSM] **19 Updating and extending Stata—Internet functionality**.) All we need to do is direct describe to look at the proper file by using the command describe using *filename*.

```
. describe using http://www.stata-press.com/data/r12/auto
Contains data                                    1978 Automobile Data
  obs:            74                             13 Apr 2011 17:45
  vars:           12
  size:        3,478

               storage   display     value
variable name    type    format      label      variable label

make           str18    %-18s                   Make and Model
price          int      %8.0gc                  Price
mpg            int      %8.0g                    Mileage (mpg)
rep78          int      %8.0g                    Repair Record 1978
headroom       float    %6.1f                    Headroom (in.)
trunk          int      %8.0g                    Trunk space (cu. ft.)
weight         int      %8.0gc                   Weight (lbs.)
length         int      %8.0g                    Length (in.)
turn           int      %8.0g                    Turn Circle (ft.)
displacement   int      %8.0g                    Displacement (cu. in.)
gear_ratio     float    %6.2f                    Gear Ratio
foreign        byte     %8.0g       origin       Car type

Sorted by:  foreign
```

This output is much more informative. There are three locations where labels are attached that help explain what the dataset contains:

1. In the first line, 1978 Automobile Data is the *data label*. It gives information about the contents of the dataset. Data can be labeled by selecting **Data > Data utilities > Label utilities > Label dataset,** or by using the label data command.

2. There is a variable label attached to each variable. Variable labels are not only how we would refer to the variable in normal, everyday conversation, but they also contain information about the units of the variables. Variables can be labeled by selecting the variable in the Variables window and editing the *Label* field in the Properties window. When doing this in the main window, be sure that the Properties window is unlocked. You can also change a variable label using the Variables Manager or by using the label variable command.

3. The foreign variable has an attached value label. Value labels allow numeric variables, such as foreign, to have words associated with numeric codes. The describe output tells you that the numeric variable foreign has value label origin associated with it. Although not revealed by describe, the variable foreign takes on the values 0 and 1, and the value label origin associates 0 with Domestic and 1 with Foreign. If you browse the data (see [GSM] **6 Using the Data Editor**), foreign appears to contain the values "Domestic" and "Foreign". The values in a variable are labeled in two stages. The value label must first be defined. This can be done in the Data Editor, or in the Variables Manager, or by selecting **Data > Data utilities > Label utilities > Manage value labels** or by typing the label define command. After the labels have been defined, they must be attached to the proper variables, either by selecting **Data > Data utilities > Label utilities > Assign value label to variables** or by using the label values command. Note: It is not necessary for the value label to have a name different from that of the variable. You could just as easily have used a value label named foreign.

Labeling datasets and variables

We will now take our `afewcars.dta` dataset and give it proper labels. We will do this with the Command window, to illustrate that it is simple to do in this fashion. Earlier in *Renaming and formatting variables* in [GSM] **6 Using the Data Editor**, we used the Data Editor to achieve a similar purpose. If you use the Data Editor for the material here, you will end up with the same commands in your log; we would like to illustrate a way to work directly with commands.

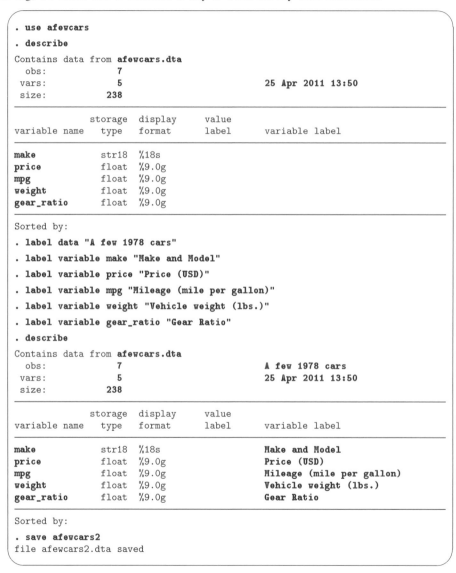

```
. use afewcars

. describe
Contains data from afewcars.dta
  obs:            7
  vars:           5                        25 Apr 2011 13:50
  size:         238

              storage   display    value
variable name type      format     label      variable label

make          str18     %18s
price         float     %9.0g
mpg           float     %9.0g
weight        float     %9.0g
gear_ratio    float     %9.0g

Sorted by:

. label data "A few 1978 cars"

. label variable make "Make and Model"

. label variable price "Price (USD)"

. label variable mpg "Mileage (mile per gallon)"

. label variable weight "Vehicle weight (lbs.)"

. label variable gear_ratio "Gear Ratio"

. describe
Contains data from afewcars.dta
  obs:            7                        A few 1978 cars
  vars:           5                        25 Apr 2011 13:50
  size:         238

              storage   display    value
variable name type      format     label      variable label

make          str18     %18s                   Make and Model
price         float     %9.0g                  Price (USD)
mpg           float     %9.0g                  Mileage (mile per gallon)
weight        float     %9.0g                  Vehicle weight (lbs.)
gear_ratio    float     %9.0g                  Gear Ratio

Sorted by:

. save afewcars2
file afewcars2.dta saved
```

Warning: When you change or define labels on a dataset in memory, it is worth saving the dataset right away. Because the actual data in the dataset did not change, Stata will not prevent you from quitting or loading a new dataset later, and you could lose your labels.

Labeling values of variables

We will now add a new indicator variable to the dataset that is 0 if the car was made in the United States and 1 if it was foreign made. Open up the Data Editor and use your previously gained knowledge to add a `foreign` variable whose values match what is shown in this listing:

```
. list, separator(0)
```

	make	price	mpg	weight	gear_r~o	foreign
1.	VW Rabbit	4697	25	1930	3.78	1
2.	Olds 98	8814	21	4060	2.41	0
3.	Chev. Monza	3667	.	2750	2.73	0
4.		4099	22	2930	3.58	0
5.	Datsun 510	5079	24	2280	3.54	1
6.	Buick Regal	5189	20	3280	2.93	0
7.	Datsun 810	8129	.	2750	3.55	1

You can create this new variable in the Data Editor if you would like to work along. (See [GSM] **6 Using the Data Editor** for help with the Data Editor.) Though the definitions of the categories "0" and "1" are clear in this context, it still would be worthwhile to give the values explicit labels, because it will make output clear to people who are not so familiar with antique automobiles. This is done with a value label.

We saw an example of creating and attaching a value label by using the point-and-click interface available in the Data Editor in *Changing data* in [GSM] **6 Using the Data Editor**. Here we will do it directly from the Command window.

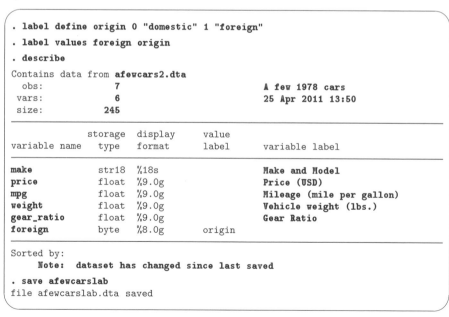

```
. label define origin 0 "domestic" 1 "foreign"
. label values foreign origin
. describe
Contains data from afewcars2.dta
  obs:            7                          A few 1978 cars
  vars:           6                          25 Apr 2011 13:50
  size:         245

              storage   display    value
variable name   type    format     label      variable label

make           str18    %18s                  Make and Model
price          float    %9.0g                 Price (USD)
mpg            float    %9.0g                 Mileage (mile per gallon)
weight         float    %9.0g                 Vehicle weight (lbs.)
gear_ratio     float    %9.0g                 Gear Ratio
foreign        byte     %8.0g      origin

Sorted by:
     Note:  dataset has changed since last saved
. save afewcarslab
file afewcarslab.dta saved
```

From this example, we can see that a value label is defined via

`label define` *labelname # "contents" # "contents"* ...

It can then be attached to a variable via

`label values` *variablename labelname*

Once again, we need to save the dataset to be sure that we do not mistakenly lose the labels later. We saved this under a new filename because we have cleaned it up, and we would like to use it in the next chapter.

If you had wanted to define the value labels by using a point-and-click interface, you could do this with the Properties window in either the Main window or in the Data Editor or by using the Variables Manager. See [GSM] **7 Using the Variables Manager** for more information.

There is more to value labels than what was covered here; see [U] **12.6.3 Value labels** for a complete treatment.

You may also add notes to your data and your variables. This feature was previously discussed in *Renaming and formatting variables* in [GSM] **6 Using the Data Editor** and *Managing notes* in [GSM] **7 Using the Variables Manager**. You can learn more about notes by typing `help notes` or you can get the full story in [D] **notes**.

10   Listing data and basic command syntax

Command syntax

This chapter gives a basic lesson on Stata's command syntax while showing how to control the appearance of a data list.

As we have seen throughout this manual, you have a choice between using menus and dialogs and using the Command window. Although many find the menus more natural and the Command window baffling at first, some practice makes working with the Command window often much faster than using menus and dialogs. The Command window can become a faster way of working because of the clean and regular syntax of Stata commands. We will cover enough to get you started; `help language` has more information and examples, and [U] **11 Language syntax** has all the details.

The syntax for the `list` command can be seen by typing `help list`:

$$\underline{\texttt{li}}\texttt{st} \; \big[\textit{varlist}\big] \; \big[\textit{if}\big] \; \big[\textit{in}\big] \; \big[\, , \textit{options}\big]$$

Here is how to read this syntax:

- Anything inside square brackets is optional. For the `list` command,
  a. *varlist* is optional. A *varlist* is a list of variable names.
  b. *if* is optional. The `if` qualifier restricts the command to run only on those observations for which the qualifier is true. We saw examples of this in [GSM] **6 Using the Data Editor**.
  c. *in* is optional. The `in` qualifier restricts the command to run on particular observation numbers.
  d. , and *options* are optional. *options* are separated from the rest of the command by a comma.
- Optional pieces do not preclude one another unless explicitly stated. For the `list` command, it is possible to use a *varlist* with *if* and *in*.
- If a part of a word is underlined, the underlined part is the minimum abbreviation. Any abbreviation at least this long is acceptable.
  a. The l in `list` is underlined, so `l`, `li`, and `lis` are all equivalent to `list`.
- Anything not inside square brackets is required. For the `list` command, only the command itself is required.

Keeping these rules in mind, let's investigate how `list` behaves when called with different arguments. We will be using the dataset `afewcars.dta` from the end of the previous chapter.

list with a variable list

Variable lists (or *varlist*s) can be specified in a variety of ways, all designed to save typing and encourage good variable names.

- The *varlist* is optional for `list`. This means that if no variables are specified, it is equivalent to specifying all variables. Another way to think of it is that the default behavior of the command is to run on all variables, unless restricted by a *varlist*.
- You can list a subset of variables explicitly, as in `list make mpg price`.
- There are also many shorthand notations:
  - m* means all variables starting with m.
  - `price-weight` means all variables from `price` through `weight` in dataset order.
  - `ma?e` means all variables starting with `ma`, followed by any character, and ending in e.

- You can list a variable by using an abbreviation unique to that variable, as in list gear_r~o. If the abbreviation is not unique, Stata returns an error message.

```
. list

              make    price    mpg    weight    gear_r~o    foreign

  1.      VW Rabbit     4697     25      1930        3.78     foreign
  2.        Olds 98     8814     21      4060        2.41    domestic
  3.    Chev. Monza     3667      .      2750        2.73    domestic
  4.                    4099     22      2930        3.58    domestic
  5.     Datsun 510     5079     24      2280        3.54     foreign

  6.    Buick Regal     5189     20      3280        2.93    domestic
  7.     Datsun 810     8129      .      2750        3.55     foreign

. l make mpg price

              make    mpg    price

  1.      VW Rabbit     25     4697
  2.        Olds 98     21     8814
  3.    Chev. Monza      .     3667
  4.                    22     4099
  5.     Datsun 510     24     5079

  6.    Buick Regal     20     5189
  7.     Datsun 810      .     8129

. list m*

              make    mpg

  1.      VW Rabbit     25
  2.        Olds 98     21
  3.    Chev. Monza      .
  4.                    22
  5.     Datsun 510     24

  6.    Buick Regal     20
  7.     Datsun 810      .

. li price-weight

     price    mpg    weight

  1.   4697     25      1930
  2.   8814     21      4060
  3.   3667      .      2750
  4.   4099     22      2930
  5.   5079     24      2280

  6.   5189     20      3280
  7.   8129      .      2750
```

```
. list ma?e

                    make

       1.     VW Rabbit
       2.        Olds 98
       3.   Chev. Monza
       4.
       5.     Datsun 510

       6.   Buick Regal
       7.   Datsun 810

. l gear_r~o

              gear_r~o

       1.        3.78
       2.        2.41
       3.        2.73
       4.        3.58
       5.        3.54

       6.        2.93
       7.        3.55
```

list with if

The if qualifier uses a logical expression to determine which observations to use. If the expression is true, the observation is used in the command; otherwise, it is skipped. The operators whose results are either true or false are

<	less than
<=	less than or equal
==	equal
>	greater than
>=	greater than or equal
!=	not equal
&	and
|	or
!	not (logical negation; ~ can also be used)
()	parentheses are for grouping to specify order of evaluation

In the logical expressions, & is evaluated before | (similar to multiplication before addition in arithmetic). You can use this in your expressions, but it is often better to use parentheses to ensure that the expressions are evaluated properly. See [U] **13.2 Operators** for the complete story.

```
. list

              make    price    mpg    weight    gear_r~o    foreign

  1.      VW Rabbit     4697     25      1930        3.78     foreign
  2.         Olds 98     8814     21      4060        2.41    domestic
  3.     Chev. Monza     3667      .      2750        2.73    domestic
  4.                    4099     22      2930        3.58    domestic
  5.      Datsun 510     5079     24      2280        3.54     foreign

  6.     Buick Regal     5189     20      3280        2.93    domestic
  7.      Datsun 810     8129      .      2750        3.55     foreign

. list if mpg > 22

              make    price    mpg    weight    gear_r~o    foreign

  1.      VW Rabbit     4697     25      1930        3.78     foreign
  3.     Chev. Monza     3667      .      2750        2.73    domestic
  5.      Datsun 510     5079     24      2280        3.54     foreign
  7.      Datsun 810     8129      .      2750        3.55     foreign

. list if (mpg > 22) & !missing(mpg)

              make    price    mpg    weight    gear_r~o    foreign

  1.      VW Rabbit     4697     25      1930        3.78     foreign
  5.      Datsun 510     5079     24      2280        3.54     foreign

. list make mpg price gear if (mpg > 22) | (price > 8000 & gear < 3.5)

              make     mpg    price    gear_r~o

  1.      VW Rabbit     25     4697       3.78
  2.         Olds 98     21     8814       2.41
  3.     Chev. Monza      .     3667       2.73
  5.      Datsun 510     24     5079       3.54
  7.      Datsun 810      .     8129       3.55

. list make mpg if mpg <= 22 in 2/4

         make     mpg

  2.    Olds 98     21
  4.               22
```

In the listings above, we see more examples of Stata treating missing numerical values as large values, as well as the care that should be taken when the `if` qualifier is applied to a variable with missing values. See [GSM] **6 Using the Data Editor**.

list with if, common mistakes

Here is a series of listings with common errors and their corrections. See if you can find the errors before reading the correct entry.

```
. list

                  make    price    mpg    weight    gear_r~o    foreign

      1.       VW Rabbit     4697     25      1930        3.78     foreign
      2.          Olds 98     8814     21      4060        2.41    domestic
      3.      Chev. Monza     3667      .      2750        2.73    domestic
      4.                      4099     22      2930        3.58    domestic
      5.      Datsun 510     5079     24      2280        3.54     foreign

      6.      Buick Regal     5189     20      3280        2.93    domestic
      7.      Datsun 810     8129      .      2750        3.55     foreign

. list if mpg=21
=exp not allowed
r(101);
```

The error arises because "equal" is expressed by ==, not by =. Corrected, it becomes

```
. list if mpg==21

              make    price    mpg    weight    gear_r~o    foreign

      2.    Olds 98     8814     21      4060        2.41    domestic
```

Other common errors with logic:

```
. list if mpg==21 if weight > 4000
invalid syntax
r(198);
. list if mpg==21 and weight > 4000
invalid 'and'
r(198);
```

Joint tests are specified with &, not with the word and or multiple ifs. The if qualifier should be if mpg==21 & weight>4000, not if mpg==21 if weight>4000. Here is its correction:

```
. list if mpg==21 & weight > 4000

              make    price    mpg    weight    gear_r~o    foreign

      2.    Olds 98     8814     21      4060        2.41    domestic
```

A problem with string variables:

```
. list if make==Datsun 510
Datsun not found
r(111);
```

Strings must be in double quotes, as in make=="Datsun 510". Without the quotes, Stata thinks that Datsun is a variable that it cannot find. Here is the correction:

```
. list if make=="Datsun 510"
```

	make	price	mpg	weight	gear_r~o	foreign
5.	Datsun 510	5079	24	2280	3.54	foreign

Confusing value labels with strings:

```
. list if foreign=="domestic"
type mismatch
r(109);
```

Value labels look like strings but the underlying variable is numeric. Variable foreign takes on values 0 and 1 but has the value label that attaches 0 to "domestic" and 1 to "foreign" (see [GSM] **9 Labeling data**). To see the underlying numeric values of variables with labeled values, use the label list command (see [D] **label**), or investigate the variable with codebook *varname*. We can correct the error here by looking for observations where foreign==0.

There is a second construction that also allows the use of the value label directly.

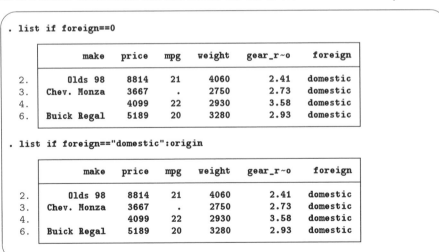

```
. list if foreign==0
```

	make	price	mpg	weight	gear_r~o	foreign
2.	Olds 98	8814	21	4060	2.41	domestic
3.	Chev. Monza	3667	.	2750	2.73	domestic
4.		4099	22	2930	3.58	domestic
6.	Buick Regal	5189	20	3280	2.93	domestic

```
. list if foreign=="domestic":origin
```

	make	price	mpg	weight	gear_r~o	foreign
2.	Olds 98	8814	21	4060	2.41	domestic
3.	Chev. Monza	3667	.	2750	2.73	domestic
4.		4099	22	2930	3.58	domestic
6.	Buick Regal	5189	20	3280	2.93	domestic

list with in

The in qualifier uses a *numlist* to give a range of observations that should be listed. *numlists* have the form of one number or *first/last*. Positive numbers count from the beginning of the dataset. Negative numbers count from the end of the dataset. Here are some examples:

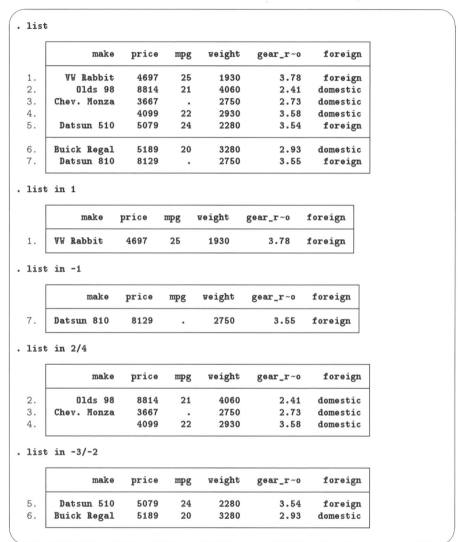

```
. list

              make    price    mpg   weight   gear_r~o    foreign

  1.      VW Rabbit     4697     25     1930       3.78    foreign
  2.         Olds 98     8814     21     4060       2.41   domestic
  3.    Chev. Monza     3667      .     2750       2.73   domestic
  4.                    4099     22     2930       3.58   domestic
  5.     Datsun 510     5079     24     2280       3.54    foreign

  6.    Buick Regal     5189     20     3280       2.93   domestic
  7.     Datsun 810     8129      .     2750       3.55    foreign

. list in 1

              make    price    mpg   weight   gear_r~o    foreign

  1.     VW Rabbit      4697     25     1930       3.78    foreign

. list in -1

              make    price    mpg   weight   gear_r~o    foreign

  7.    Datsun 810      8129      .     2750       3.55    foreign

. list in 2/4

              make    price    mpg   weight   gear_r~o    foreign

  2.         Olds 98     8814     21     4060       2.41   domestic
  3.    Chev. Monza     3667      .     2750       2.73   domestic
  4.                    4099     22     2930       3.58   domestic

. list in -3/-2

              make    price    mpg   weight   gear_r~o    foreign

  5.    Datsun 510      5079     24     2280       3.54    foreign
  6.    Buick Regal     5189     20     3280       2.93   domestic
```

Controlling the list output

The fine control over list output is exercised by specifying one or more options. You can use sepby() to separate observations by variable. abbreviate() specifies the minimum number of characters to abbreviate a variable in the output. divider draws a vertical line between the variables in the list.

```
. sort foreign
. list ma p g f, sepby(foreign)
```

	make	price	gear_r~o	foreign
1.		4099	3.58	domestic
2.	Olds 98	8814	2.41	domestic
3.	Buick Regal	5189	2.93	domestic
4.	Chev. Monza	3667	2.73	domestic
5.	Datsun 510	5079	3.54	foreign
6.	Datsun 810	8129	3.55	foreign
7.	VW Rabbit	4697	3.78	foreign

```
. list make weight gear, abbreviate(10)
```

	make	weight	gear_ratio
1.		2930	3.58
2.	Olds 98	4060	2.41
3.	Buick Regal	3280	2.93
4.	Chev. Monza	2750	2.73
5.	Datsun 510	2280	3.54
6.	Datsun 810	2750	3.55
7.	VW Rabbit	1930	3.78

```
. list, divider
```

	make	price	mpg	weight	gear_r~o	foreign
1.		4099	22	2930	3.58	domestic
2.	Olds 98	8814	21	4060	2.41	domestic
3.	Buick Regal	5189	20	3280	2.93	domestic
4.	Chev. Monza	3667	.	2750	2.73	domestic
5.	Datsun 510	5079	24	2280	3.54	foreign
6.	Datsun 810	8129	.	2750	3.55	foreign
7.	VW Rabbit	4697	25	1930	3.78	foreign

The separator() option draws a horizontal line at specified intervals. When not specified, it defaults to a value of 5.

```
. list, separator(3)

               make   price   mpg   weight   gear_r~o   foreign

  1.                    4099    22     2930       3.58   domestic
  2.           Olds 98   8814    21     4060       2.41   domestic
  3.      Buick Regal    5189    20     3280       2.93   domestic

  4.      Chev. Monza    3667     .     2750       2.73   domestic
  5.      Datsun 510     5079    24     2280       3.54    foreign
  6.      Datsun 810     8129     .     2750       3.55    foreign

  7.        VW Rabbit    4697    25     1930       3.78    foreign
```

More

When you see a —more— prompt at the bottom of the Results window, it means that there is more information to be displayed. This happens, for example, when you are listing many observations.

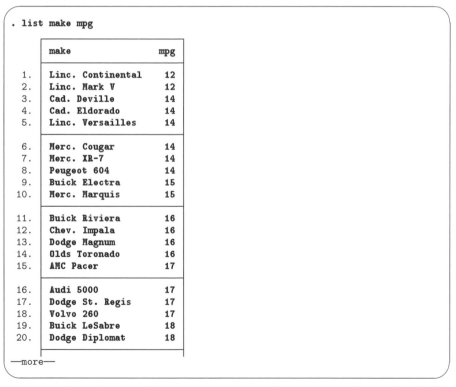

```
. list make mpg

               make             mpg

  1.   Linc. Continental        12
  2.   Linc. Mark V             12
  3.   Cad. Deville             14
  4.   Cad. Eldorado            14
  5.   Linc. Versailles         14

  6.   Merc. Cougar             14
  7.   Merc. XR-7               14
  8.   Peugeot 604              14
  9.   Buick Electra            15
 10.   Merc. Marquis            15

 11.   Buick Riviera            16
 12.   Chev. Impala             16
 13.   Dodge Magnum             16
 14.   Olds Toronado            16
 15.   AMC Pacer                17

 16.   Audi 5000                17
 17.   Dodge St. Regis          17
 18.   Volvo 260                17
 19.   Buick LeSabre            18
 20.   Dodge Diplomat           18

—more—
```

If you want to see the next screen of text, you have a few options: press any key, such as the Spacebar; click on the **More** button, 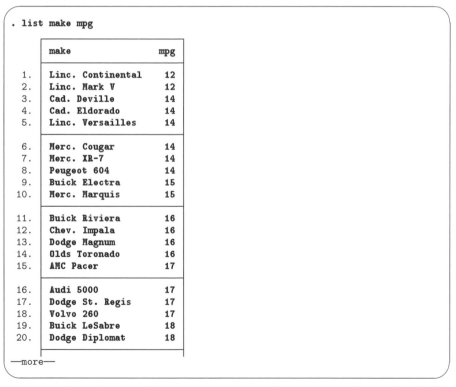; or click on the blue —more— at the bottom of the Results window. To see just the next line of text, press *Return*.

Break

If you want to interrupt a Stata command, click on the **Break** button, 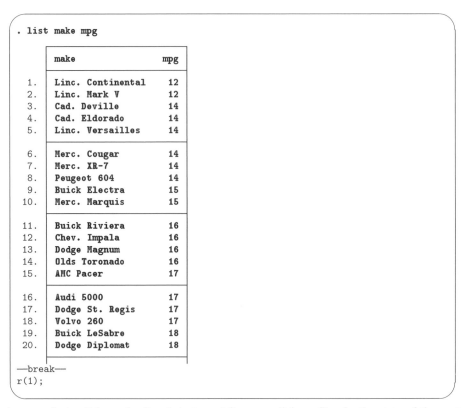. If you see a —more— prompt at the bottom of the Results window and wish to interrupt it, click on the **Break** button or press *q*.

```
. list make mpg

     |  make               mpg  |
     |---------------------------|
  1. |  Linc. Continental   12  |
  2. |  Linc. Mark V        12  |
  3. |  Cad. Deville        14  |
  4. |  Cad. Eldorado       14  |
  5. |  Linc. Versailles    14  |
     |---------------------------|
  6. |  Merc. Cougar        14  |
  7. |  Merc. XR-7          14  |
  8. |  Peugeot 604         14  |
  9. |  Buick Electra       15  |
 10. |  Merc. Marquis       15  |
     |---------------------------|
 11. |  Buick Riviera       16  |
 12. |  Chev. Impala        16  |
 13. |  Dodge Magnum        16  |
 14. |  Olds Toronado       16  |
 15. |  AMC Pacer           17  |
     |---------------------------|
 16. |  Audi 5000           17  |
 17. |  Dodge St. Regis     17  |
 18. |  Volvo 260           17  |
 19. |  Buick LeSabre       18  |
 20. |  Dodge Diplomat      18  |
     |---------------------------|
—break—
r(1);
```

It is always safe to click on the **Break** button. After you click on **Break**, the state of the system is the same as if you had never issued the original command.

11 Creating new variables

generate and replace

This chapter shows the basics of creating and modifying variables in Stata. We saw how to work with the Data Editor in [GSM] **6 Using the Data Editor**—this chapter shows how we would do this from the Command window. The two primary commands used for this are

- `generate` for creating new variables. It has a minimum abbreviation of g.
- `replace` for replacing the values of an existing variable. It may not be abbreviated because it alters existing data and hence can be considered dangerous.

The most basic form for creating new variables is `generate` *newvar* = *exp*, where *exp* is any kind of *expression*. Both `generate` and `replace` can be used with `if` and `in` qualifiers, of course. An expression is a formula made up of constants, existing variables, operators, and functions. Some examples of expressions (using variables from the `auto` dataset) would be `2 + price`, `weight^2`, or `sqrt(gear_ratio)`.

The operators defined in Stata are given in the table below:

	Arithmetic		Logical		Relational (numeric and string)
+	addition	!	not	>	greater than
−	subtraction	|	or	<	less than
*	multiplication	&	and	>=	> or equal
/	division			<=	< or equal
^	power			==	equal
				!=	not equal
+	string concatenation				

Stata has many mathematical, statistical, string, date, time-series, and programming functions. See `help functions` for the basics, and see [D] **functions** for a complete list and full details of all the built-in functions.

You can use menus and dialogs to create new variables and modify existing variables by selecting menu items from the **Data > Create or change data** menu. This feature can be handy for finding functions quickly. We will use the Command window for the examples in this chapter, however, because we would like to illustrate simple usage and some pitfalls.

generate

There are some details you should know about the `generate` command:

- The basic form of the `generate` command is `generate` *newvar* = *exp*, where *newvar* is a new variable name and *exp* is any valid expression. You will get an error message if you try to `generate` a variable that already exists.
- An algebraic calculation using a missing value yields a missing value, as does division by zero, etc.

- If missing values are generated, the number of missing values in *newvar* is always reported. If Stata says nothing about missing values, then no missing values were generated.
- You can use `generate` to set the storage type of the new variable as it is generated. You might want to create an indicator variable as a `byte`, for example, because it saves 3 bytes per observation over using the default storage type of `float`.

Below are some examples of creating new variables from the `afewcarslab` dataset that we created in *Labeling values of variables* in [GSM] **9 Labeling data**. Some examples are nonsensical—they are only for illustrating the `generate` command. The last example shows a way to generate an indicator variable for cars weighing more than 3,000 pounds. Logical expressions in Stata result in 1 for "true" and 0 for "false." The `if` qualifier is needed to be sure that the computations are done only for observations where `weight` is not missing.

```
. use afewcarslab
(A few 1978 cars)

. list make mpg weight

            make    mpg   weight

  1.    VW Rabbit     25     1930
  2.      Olds 98     21     4060
  3.   Chev. Monza      .     2750
  4.                   22     2930
  5.    Datsun 510     24     2280

  6.   Buick Regal     20     3280
  7.    Datsun 810      .     2750

. generate lphk = 3.7854 * (100 / 1.6093) / mpg
(2 missing values generated)

. label var lphk "Liters per 100km"

. g strange = sqrt(mpg*weight)
(2 missing values generated)

. gen huge = weight >= 3000 if !missing(weight)

. l make mpg weight lphk strange huge

            make    mpg   weight        lphk     strange   huge

  1.    VW Rabbit     25     1930    9.408812    219.6588      0
  2.      Olds 98     21     4060    11.20097    291.9932      1
  3.   Chev. Monza      .     2750           .           .      0
  4.                   22     2930    10.69183    253.8897      0
  5.    Datsun 510     24     2280    9.800845    233.9231      0

  6.   Buick Regal     20     3280    11.76101     256.125      1
  7.    Datsun 810      .     2750           .           .      0
```

replace

Whereas `generate` is used to create new variables, `replace` is the command used for existing variables. Stata uses two different commands to prevent you from accidentally modifying your data. The `replace` command cannot be abbreviated. Stata generally requires you to spell out completely any command that can alter your existing data.

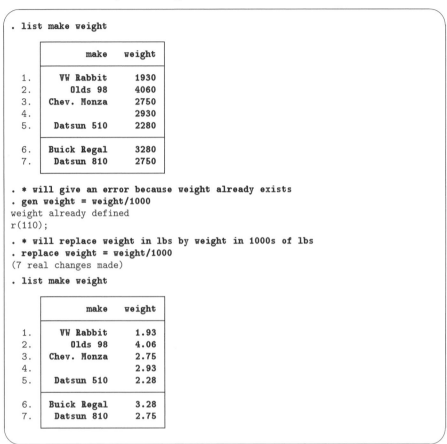

```
. list make weight

          make     weight

  1.    VW Rabbit      1930
  2.      Olds 98      4060
  3.  Chev. Monza      2750
  4.                   2930
  5.   Datsun 510      2280

  6.  Buick Regal      3280
  7.   Datsun 810      2750

. * will give an error because weight already exists
. gen weight = weight/1000
weight already defined
r(110);
. * will replace weight in lbs by weight in 1000s of lbs
. replace weight = weight/1000
(7 real changes made)
. list make weight

          make     weight

  1.    VW Rabbit      1.93
  2.      Olds 98      4.06
  3.  Chev. Monza      2.75
  4.                   2.93
  5.   Datsun 510      2.28

  6.  Buick Regal      3.28
  7.   Datsun 810      2.75
```

Suppose that you want to create a new variable, `predprice`, which will be the predicted price of the cars in the following year. You estimate that domestic cars will increase in price by 5% and foreign cars, by 10%.

One way to create the variable would be to first use `generate` to compute the predicted domestic car prices. Then use `replace` to change the missing values for the foreign cars to their proper values.

```
. gen predprice = 1.05*price if foreign==0
(3 missing values generated)
. replace predprice = 1.10*price if foreign==1
(3 real changes made)
. list make foreign price predprice, nolabel
```

	make	foreign	price	predpr~e
1.	VW Rabbit	1	4697	5166.7
2.	Olds 98	0	8814	9254.7
3.	Chev. Monza	0	3667	3850.35
4.		0	4099	4303.95
5.	Datsun 510	1	5079	5586.9
6.	Buick Regal	0	5189	5448.45
7.	Datsun 810	1	8129	8941.9

Of course, because foreign is an indicator variable, we could generate the predicted variable with one command:

```
. gen predprice2 = (1.05 + 0.05*foreign)*price
. list make foreign price predprice predprice2, nolabel
```

	make	foreign	price	predpr~e	predpr~2
1.	VW Rabbit	1	4697	5166.7	5166.7
2.	Olds 98	0	8814	9254.7	9254.7
3.	Chev. Monza	0	3667	3850.35	3850.35
4.		0	4099	4303.95	4303.95
5.	Datsun 510	1	5079	5586.9	5586.9
6.	Buick Regal	0	5189	5448.45	5448.45
7.	Datsun 810	1	8129	8941.9	8941.9

generate with string variables

Stata is smart. When you generate a variable and the expression evaluates to a string, Stata creates a string variable with a storage type as long as necessary, and no longer than that. `where` is a `str1` in the following example:

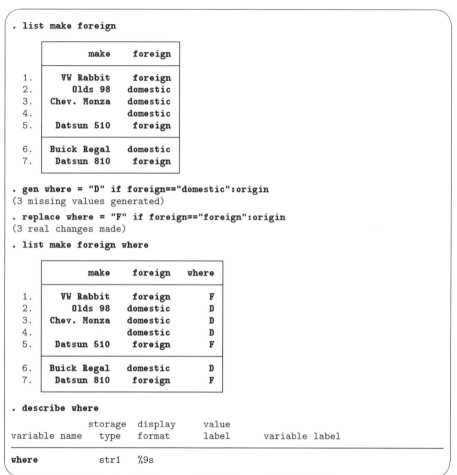

```
. list make foreign

              make     foreign

   1.     VW Rabbit     foreign
   2.       Olds 98    domestic
   3.   Chev. Monza    domestic
   4.                  domestic
   5.    Datsun 510     foreign

   6.   Buick Regal    domestic
   7.    Datsun 810     foreign

. gen where = "D" if foreign=="domestic":origin
(3 missing values generated)
. replace where = "F" if foreign=="foreign":origin
(3 real changes made)
. list make foreign where

              make     foreign    where

   1.     VW Rabbit     foreign        F
   2.       Olds 98    domestic        D
   3.   Chev. Monza    domestic        D
   4.                  domestic        D
   5.    Datsun 510     foreign        F

   6.   Buick Regal    domestic        D
   7.    Datsun 810     foreign        F

. describe where

               storage   display     value
variable name    type     format     label       variable label
_____
where             str1     %9s
```

Stata has some useful tools for working with string variables. Here we split the `make` variable into make and model, and then create a variable that has the model together with where the model was manufactured:

```
. gen model = substr(make, strpos(make," ")+1, .)
(1 missing value generated)

. gen modelwhere = model + " " + where

. list make where model modelwhere
```

	make	where	model	modelw~e
1.	VW Rabbit	F	Rabbit	Rabbit F
2.	Olds 98	D	98	98 D
3.	Chev. Monza	D	Monza	Monza D
4.		D		D
5.	Datsun 510	F	510	510 F
6.	Buick Regal	D	Regal	Regal D
7.	Datsun 810	F	810	810 F

There are a few things to note about how these commands work:

1. strpos(s_1, s_2) produces an integer equal to the first position in the string s_1 at which the string s_2 is found, or 0 if it is not found. In this example, strpos(make," ") finds the position of the first space in each observation of make.

2. substr($s, start, len$) produces a string of length len beginning at character $start$ of string s. If $c_1 = .$, the result is the string from character $start$ to the end of string s.

3. Putting 1 and 2 together: substr(s,strpos(s," ")+1,.) will always give the string s with its first word removed. Because make contains both the make and model of each car, and make never contains a space in this dataset, we have found each car's model.

4. The operator "+", when applied to string variables, will concatenate the strings (that is, join them together). The expression "this" + "that" results in the string "thisthat". When the variable modelwhere was generated, a space (" ") was added between the two strings.

5. The missing value for a string is nothing special—it is simply the empty string "". Thus the value of modelwhere for the car with no make or model is " D" (note the leading space).

12 Deleting variables and observations

clear, drop, and keep

In this chapter, we will present the tools for paring observations and variables from a dataset. We saw how to do this using the Data Editor in [GSM] **6 Using the Data Editor**; this chapter presents the methods for doing so from the Command window.

There are three main commands for removing data and other Stata objects from memory: `clear`, `drop`, and `keep`. Remember that they affect only what is in memory. None of these commands alters anything that has been saved to disk.

clear and drop _all

Suppose that you are working on an analysis or a simulation and you need to clear out Stata's memory so that you can impute different values or simulate a new dataset. You are not interested in saving any of the changes you have made to the dataset in memory—you would just like to have an empty dataset. What you do depends on how much you want to clear out: at any time, you can have not only data but also metadata such as value labels, saved results from previous commands, and saved matrices. The `clear` command will let you carefully clear out data or other objects; we are interested only in simple usage here. For more information, see `help clear` and [D] **clear**.

If you type the command `clear` into the Command window, it will remove all variables and value labels. In basic usage, this is typically enough. It has the nice property that it does not remove any saved results, so you can load a new dataset and predict values by using saved estimation results from a model fit on a previous dataset. See `help postest` and [U] **20 Estimation and postestimation commands** for more information.

If you want to be sure that everything is cleared out, use the command `clear all`. This command will clear Stata's memory of data and all auxiliary objects so that you can start with a clean slate. The first time you use `clear all` while you have a graph or dialog open, you may be surprised when that graph or dialog closes; this is necessary so that Stata can free all memory that is being used.

If you want to get rid of just the data and nothing else, you can use the command `drop _all`.

drop

The `drop` command is used to remove variables or observations from the dataset in memory.
- If you want to drop variables, use `drop` *varlist*.
- If you want to drop observations, use `drop` with an `if` or an `in` qualifier, or both.

We will use the afewcarslab dataset, as usual:

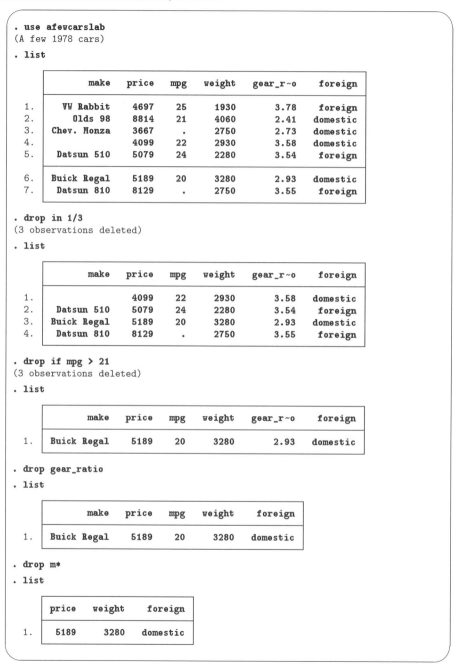

```
. use afewcarslab
(A few 1978 cars)

. list
```

	make	price	mpg	weight	gear_r~o	foreign
1.	VW Rabbit	4697	25	1930	3.78	foreign
2.	Olds 98	8814	21	4060	2.41	domestic
3.	Chev. Monza	3667	.	2750	2.73	domestic
4.		4099	22	2930	3.58	domestic
5.	Datsun 510	5079	24	2280	3.54	foreign
6.	Buick Regal	5189	20	3280	2.93	domestic
7.	Datsun 810	8129	.	2750	3.55	foreign

```
. drop in 1/3
(3 observations deleted)

. list
```

	make	price	mpg	weight	gear_r~o	foreign
1.		4099	22	2930	3.58	domestic
2.	Datsun 510	5079	24	2280	3.54	foreign
3.	Buick Regal	5189	20	3280	2.93	domestic
4.	Datsun 810	8129	.	2750	3.55	foreign

```
. drop if mpg > 21
(3 observations deleted)

. list
```

	make	price	mpg	weight	gear_r~o	foreign
1.	Buick Regal	5189	20	3280	2.93	domestic

```
. drop gear_ratio

. list
```

	make	price	mpg	weight	foreign
1.	Buick Regal	5189	20	3280	domestic

```
. drop m*

. list
```

	price	weight	foreign
1.	5189	3280	domestic

These changes are only to the data in memory. If you wanted to make the changes permanent, you would need to save the dataset.

keep

keep tells Stata to drop all variables except those specified explicitly or through the use of an *if* or *in* expression. Just like drop, keep can be used with a *varlist* or with qualifiers, but not with both at once. We use a clear command at the start of this example so that we can reload the afewcarslab dataset:

```
. clear
. use afewcarslab
(A few 1978 cars)
. list
```

	make	price	mpg	weight	gear_r~o	foreign
1.	VW Rabbit	4697	25	1930	3.78	foreign
2.	Olds 98	8814	21	4060	2.41	domestic
3.	Chev. Monza	3667	.	2750	2.73	domestic
4.		4099	22	2930	3.58	domestic
5.	Datsun 510	5079	24	2280	3.54	foreign
6.	Buick Regal	5189	20	3280	2.93	domestic
7.	Datsun 810	8129	.	2750	3.55	foreign

```
. keep in 4/7
(3 observations deleted)
. list
```

	make	price	mpg	weight	gear_r~o	foreign
1.		4099	22	2930	3.58	domestic
2.	Datsun 510	5079	24	2280	3.54	foreign
3.	Buick Regal	5189	20	3280	2.93	domestic
4.	Datsun 810	8129	.	2750	3.55	foreign

```
. keep if mpg <= 21
(3 observations deleted)
. list
```

	make	price	mpg	weight	gear_r~o	foreign
1.	Buick Regal	5189	20	3280	2.93	domestic

```
. keep m*
. list
```

	make	mpg
1.	Buick Regal	20

Notes

13 Using the Do-file Editor—automating Stata

The Do-file Editor

Stata comes with an integrated text editor called the Do-file Editor, which can be used for many tasks. It gets its name from the term *do-file*, which is a file containing a list of commands for Stata to run (called a batch file or a script in other settings). See [U] **16 Do-files** for more information. Although the Do-file Editor has advanced features that can help in writing such files, it can also be used to build up a series of commands that can then be submitted to Stata all at once. This feature can be handy when writing a loop to process multiple variables in a similar fashion or when doing complex, repetitive tasks interactively.

To get the most from this chapter, you should work through it at your computer. Start by opening the do-file editor, either by clicking on the **Do-file Editor** button ![icon] or by typing doedit in the Command window and pressing *Return*.

The Do-file Editor toolbar

The Do-file Editor has seven buttons. Many of the buttons share a similar purpose with their look-alikes in the main Stata toolbar.

If you ever forget what a button does, hover the mouse pointer over a button for a moment, and a tooltip will appear with a description of that button.

Open: Open a do-file from disk in a new tab in the Do-file Editor.

Save: Save the current do-file to disk.

Print: Print the contents of the Do-file Editor.

Find: Open the *Find* dialog for finding text.

Show: Toggle display of invisible characters.

Execute Quietly (run): Run the commands in the do-file without showing any output. If text is highlighted, the button becomes the **Execute Selection Quietly (run)** button and will run just the selected lines, suppressing all output. This feature is useful when assembling long series of commands. We will refer to this as the **Run** button.

Execute (do): Run the commands in the do-file, showing all commands and their output. If text is highlighted, the button becomes the **Execute Selection (do)** button and will run only the selected lines, showing all output. We will refer to this as the **Do** button.

Using the Do-file Editor

Suppose that we would like to analyze fuel usage for 1978 automobiles in a manner similar to what we did in [GSM] **1 Introducing Stata—sample session**. We know that we will be issuing many commands to Stata during our analysis and that we want to be able to reproduce our work later without having to type each command again.

We can do this easily in Stata: simply save a text file containing the commands. When that is done, we can tell Stata to run the file and execute each command in sequence. Such a file is known as a Stata *do-file*; see [U] **16 Do-files**.

To analyze fuel usage of 1978 automobiles, we would like to create a new variable giving gallons per mile. We would like to see how that variable changes in relation to vehicle weight for both domestic and imported cars. Performing a regression with our new variable would be a good first step.

To get started, click on the **Do-file Editor** button to open the Do-file Editor. After the Do-file Editor opens, type the commands below into the Do-file Editor. Purposely misspell the name of the `foreign` variable on the fifth line. (We are intentionally making some common mistakes and then pointing you to the solutions. This will save you time later.)

```
* an example do-file
sysuse auto
generate gp100m = 100/mpg
label var gp100m "Gallons per 100 miles"
regress gp100m weight foreing
```

Here is what your Do-file Editor should look like now:

You will notice that the color of the text changes as you type. The different colors are examples of the Do-file Editor's *syntax highlighting*. The colors and text properties of the syntax elements can be changed by right-clicking in the Do-file Editor, selecting **Preferences...**, and then clicking on the **Syntax Highlighting** tab in the resulting window. Click on the **Do** button, , to run the commands. When you click on the **Do** button, Stata executes the commands in sequence, and the results appear in the Results window:

```
. do /tmp/SD00001.000000

. * an example do-file

. sysuse auto
(1978 Automobile Data)

. generate gp100m = 100/mpg

. label var gp100m "Gallons per 100 miles"

. regress gp100m weight foreing
variable foreing not found
r(111);

.
end of do-file
```

The do "/tmp/..." command is how Stata executes the commands in the Do-file Editor. Stata saves the commands to a temporary file and issues the do command to execute them. Everything worked as planned until Stata saw the misspelled variable. The first three commands were executed, but an error was produced on the fourth. Stata does not know of a variable named foreing. We need to go back to the Do-file Editor and change the misspelled variable name to foreign in the last line:

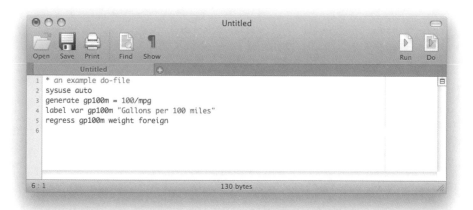

We click on the **Do** button again. Alas, Stata now fails on the first line—it will not overwrite the dataset in memory that we had changed.

```
. do /tmp/SD00001.000000

. * an example do-file

. sysuse auto
no; data in memory would be lost
r(4);

.
end of do-file
```

We now have a choice for what we should do:

- We can put a clear command in our do-file as the very first command. This automatically clears out Stata's memory before the do-file tries to load the auto dataset. This is convenient, but dangerous, because it defeats Stata's protection against throwing away changes without warning.

- We can type a `clear` command to manually clear the dataset and then process the do-file again. This process can be aggravating when building a complicated do-file.

Here is some advice: Automatically clear Stata's memory while debugging the do-file. Once the do-file is in its final form, decide the context in which it will be used. If it will be used in a highly automated environment (such as when certifying), the do-file should still automatically clear Stata's memory. If it will be used rarely, do not clear Stata's memory. This decision will save much heartache. We will add a `clear` option to the `sysuse` command to automatically clear the dataset in Stata's memory before the do-file runs:

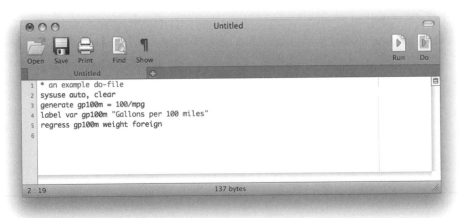

The do-file now runs well, as clicking on the **Do** button shows:

```
. do /tmp/SD00001.000000

. * an example do-file
. sysuse auto, clear
(1978 Automobile Data)

. generate gp100m = 100/mpg

. label var gp100m "Gallons per 100 miles"

. regress gp100m weight foreign
```

Source	SS	df	MS
Model	91.1761694	2	45.5880847
Residual	28.4000913	71	.400001287
Total	119.576261	73	1.63803097

Number of obs =	74
F(2, 71) =	113.97
Prob > F =	0.0000
R-squared =	0.7625
Adj R-squared =	0.7558
Root MSE =	.63246

gp100m	Coef.	Std. Err.	t	P>|t|	[95% Conf. Interval]	
weight	.0016254	.0001183	13.74	0.000	.0013896	.0018612
foreign	.6220535	.1997381	3.11	0.003	.2237871	1.02032
_cons	-.0734839	.4019932	-0.18	0.855	-.8750354	.7280677

```
.
end of do-file
```

You might want to select **File > Save As...** to save this do-file while the Do-file Editor is in front. Later, you could select **File > Open...** to open it and then add more commands as you move forward with your analysis. By saving the commands of your analysis in a do-file as you go, you do not have to worry about retyping them with each new Stata session. Think hard about removing the `clear` option from the first command.

After you have saved your do-file, you can execute the commands it contains by typing do *filename*, where the *filename* is the name of your do-file.

The File menu

When the Do-file Editor is in front, the **File** menu's items apply to do-files. You may choose any of these menu items: create a **New Do-file**, **Open...** an existing file, **Save** the current file, save the current file under a new name with **Save As...**, or **Print...** the current file. There are also buttons on the Do-file Editor's toolbar that correspond to these features.

You can also select **File > Insert File...** to insert the contents of another file at the current cursor position in the Do-file Editor.

Editing tools

The **Edit** menu of the Do-file Editor includes the standard **Undo**, **Redo**, **Cut**, **Copy**, and **Paste** capabilities. There are several other **Edit** menu features that you might find useful:

- You may select the current line with **Select Line**.
- You may delete the current line with **Delete Line**.
- **Find > Go To Line...** will allow you to jump to a specific line number. The line numbers are displayed at the left of the Do-file Editor window.
- **Advanced** leads to a submenu with some programmer's friends:
    - **Shift Right** indents the selection by one tab.
    - **Shift Left** unindents the selection by one tab.
    - **Make Selection Uppercase** converts the selection to all capital letters.
    - **Make Selection Lowercase** converts the selection to all lowercase letters.
    - **Show Nonprinting Characters** displays space, tab and newline (paragraph) characters.
    - **Wrap Line** wraps long lines while keeping the proper line numbers.

Matching and balancing of parentheses ( ), braces { }, and brackets [ ] are also available from the **Edit** menu. When you select **Edit > Find > Match Brace**, the Do-file Editor looks at the character immediately to the right of the cursor. If it is one of the characters that the editor can match, the editor will find the matching character and place the cursor immediately in front of it. If there is no match, you will hear a beep and the cursor will not move.

When you select **Edit > Find > Balance Braces**, the Do-file Editor looks to the left and right of the current cursor position or selection and creates a selection that includes the narrowest level of matching characters. If you select **Balance Braces** again, the editor will expand the selection to include the next level of matching characters. If there is no match, you will hear a beep and the cursor will not move. Balancing braces is useful for working with complicated expressions or blocks of code defined by loops or `if` commmands. See [P] **foreach**, [P] **forvalues**, [P] **while**, and [P] **if** for more information.

Balance Braces is easier to explain with an example. Type (now (is the) time) in the Do-file Editor. Place the cursor between the words `is` and `the`. Select **Edit > Find > Balance Braces**. The Do-file Editor will select (`is the`). If you select **Balance Braces** again, the Do-file Editor will select (`now (is the) time`).

Editing tip: You can split the Do-file Editor window horizontally by clicking on the **Split Window** button ⊟. This is useful for looking at two widely separated places in a file at the same time. To go back to a single pane view, click on the close button in either of the split panes.

The View > Do-file Editor menu

You have already learned about the **Do** button. Selecting **View > Do-file Editor > Execute (do)** is equivalent to clicking on the **Execute (do)** button.

Selecting **View > Do-file Editor > Execute (do) to Bottom** will send all the commands from the current line through the end of the contents of the Do-file Editor to the Command window. This method is a quick way to run a part of a do-file.

You have also already learned about the **Run** button. Selecting **View > Do-file Editor > Execute Quietly (run)** is equivalent to clicking on the **Execute Quietly (run)** button. Executing quietly still executes all the commands in the Do-file Editor, but does not produce any output. It is unlikely that you will ever need to use the **Run** button; see [U] **16.4.2 Suppressing output** for more details.

Do and **Run** are equivalent to Stata's do and run commands; see [U] **16 Do-files** for a complete discussion.

You can also preview files in the Viewer by selecting **View > Do-file Editor > Show File in Viewer**. This feature is useful when working with files that use Stata's SMCL tags, such as when writing help files or editing log files.

Saving interactive commands from Stata as a do-file

While working interactively with Stata, you might decide that you would like to rerun the last several commands that you typed interactively. From the Review window, you can send highlighted commands or even the entire contents to the Do-file Editor. You can also save commands as a do-file and open that file in the Do-file Editor. You can copy a command from a dialog (rather than submit it) and paste it into the Do-file Editor. See [GSM] **6 Using the Data Editor** for details. Also see [R] **log** for information on the cmdlog command, which allows you to log all commands that you type in Stata to a do-file.

14 Graphing data

Working with graphs

Stata has a rich system for graphical representation of data. The main command for creating graphs is unsurprisingly named `graph`. Behind this plain name is a wealth of tools. In this chapter, we will make one simple graph to point out the basics of the Graph window. See the *Graphics Reference Manual* for more information about all aspects of working with graphs.

A simple graph example

In the sample session of [GSM] **1 Introducing Stata—sample session**, we made a scatterplot, added a fitted regression line, and made a grid of scatterplots to allow comparisons across groups. Here we make a simple box plot that shows the displacements of the cars' engines and how they compare across repair records within the place of manufacture of the cars, using the `auto` dataset (`sysuse auto`).

We select **Graphics > Box plot**, choose `displacement` in the *Variables* field on the **Main** tab, click on the **Categories** tab, check the *Group 1* checkbox and enter `rep78` for the first grouping variable, and check the *Group 2* checkbox and enter `foreign` for the second grouping variable. Finally, we click on the **Submit** button so that we could easily make changes to the graph if need be. After we look at the graph, we realize that we forgot the title. We close the Graph window, click on the **Titles** tab of the *graph box* dialog, type the title `Displacement across Repairs within Origin`, and click on the **Submit** button again.

The Graph window comes up, showing us the fruit of our labor:

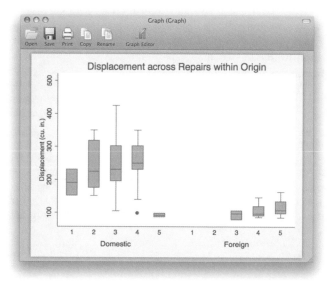

Graph window

When the Graph window comes up, it shows our graph in a window with a toolbar. The first four buttons are familiar to us from other Stata windows: **Open**, **Save**, **Print**, and **Copy**. The next two buttons are new:

Rename Graph: This button allows the graph to be renamed. Why would you do this? If you would like to have multiple graphs open at once, the graphs need to be named. So, you can click on the **Rename** button to give a graph a name. This window will then remain open when you create your next graph.

Graph Editor: Stata has a Graph Editor that allows you to manipulate and edit your graph. This feature will be introduced in the next chapter.

The inactive buttons to the right of the **Graph Editor** button are used by the Graph Editor, so their meanings will become clear in the next chapter.

We decide that we like this graph and would like to save it. We can save it either by clicking on the **Save** button and choosing a name and a location or by right-clicking on the Graph window itself, and selecting **Save As...**.

Saving and printing graphs

You can save a graph once it is displayed by right-clicking on its window and selecting **Save As...**. You can print a graph by right-clicking on its window and selecting **Print...**. You can also use the **File** menu to save or print a graph.

Right-clicking on the Graph window

Right-clicking on the Graph window displays a menu from which you can select the following:

- **Save As...** to save the graph to disk.
- **Copy** to copy the graph to the Clipboard.
- **Start Graph Editor** to start the Graph Editor.
- **Preferences...** to edit the preferences for graphs.
- **Print...** to print the contents of the Graph window.

The Graph button

The **Graph** button, , is located on the main toolbar. Clicking on the button brings the topmost Graph window to the front of all other windows. Hold down the button to select a Graph window to bring it to the front of all other windows. If you ever decide to close the Graph window, you can reopen it only by reissuing a Stata command that draws a new graph.

15 Editing graphs

Working with the Graph Editor

With Stata's Graph Editor, you can change almost anything on your graph, and you can add text, lines, arrows, and markers wherever you would like.

We will first make a graph to edit and will then point out the tools in the Graph Editor. Here is the command that we will use to make the graph:

```
. scatter mpg weight, name(mygraph) title(Mileage vs. Vehicle Weight)
```

Start the Editor by right-clicking on your graph and selecting **Start Graph Editor**. Click once on the title of the graph. Here is a picture of the Graph Editor with its elements labeled.

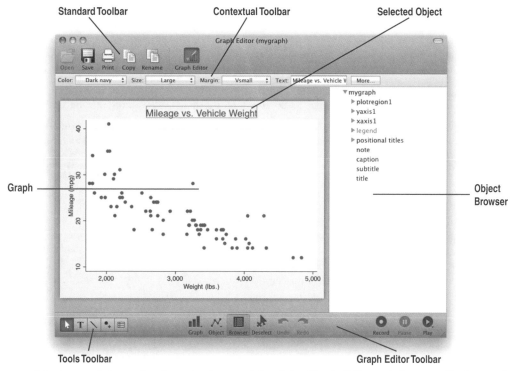

Select any of the tools along the bottom left of the Graph Editor window to edit the graph. The Pointer (Select Tool), ▮, is selected by default.

You can change the properties of objects or drag them to new locations by using the Pointer. As you select objects with the Pointer, a Contextual Toolbar will appear just above the graph. In our example, the title of the graph is selected, so the Contextual Toolbar has controls that are relevant for editing titles. You can use any of the controls on the Contextual Toolbar to immediately change the most important properties of the selected object. Right-click on an object to access more properties and operations. Hold the *Shift* key when dragging objects to constrain the movement to horizontal, vertical, or 90-degree angles.

Add text, lines, or markers (with optional labels) to your graph by using the three *Add...* tools—
T, , and . Lines can be changed to arrows by using the Contextual Toolbar. If you do not like the default properties, simply change their settings in the Contextual Toolbar before adding the text, line, or marker. The new setting will then be applied to all added objects, even in future Stata sessions.

Do not be afraid to try things. If you do not like a result, change it back by using the same tool or by clicking on the **Undo** button, , in the Graph Editor toolbar that runs along the bottom of the window. **Edit > Undo** in the main menu does the same thing.

Remember to reselect the Pointer tool when you want to drag objects or change their properties.

You can move objects on the graph and have the rest of the objects adjust their position to accommodate the move with the Grid Edit tool, . With this tool, you are repositioning objects in the underlying grid that holds the objects in the graph. Some graphs, for example, by graphs, are composed of nested grids. You can reposition objects only within the grid that contains them; they cannot be moved to other grids.

You can also select objects in the Object Browser along the right of the graph. This window shows a hierarchical listing of the objects in the graph. Clicking or right-clicking on an object in the Object Browser is the same as clicking or right-clicking on the object in the graph.

The Graph Editor has the ability to record your actions and play them back on later graphs. When you click on the **Record** button, , every editing action you take, including undos and redos, is recorded. If you would like to do some editing that is not recorded, you can click on the **Pause** button, . You can click on the **Pause** button again to resume recording. When you are done with your recording, click on the **Record** button. You will be prompted to save your recording. Any recording you save is available from the **Play Recording** button, , and may be applied to future graphs. You can even play a recording in any Stata graph command by using the play option. See *Graph Recorder* in [G-1] **graph editor** for more information.

Stop the editor by selecting **File > Stop Graph Editor** from the main menu or by clicking on the **Graph Editor** button. When you stop the Graph Editor, you will be prompted to save your graph if you have made any changes. If you do not save your graph, your changes will not be lost, but you will risk losing them if you create a new graph in the same Graph window. You must stop the Editor if you would like to work on other tasks in Stata.

Here are a few of the things that you can do with the Editor:

- Add annotations using lines, arrows, and text.
- Add or remove grid lines or reference lines.
- Add or modify titles, captions, and notes.
- Change scatterplots to line plots, connected plots, areas, bars, spikes, or drop lines—and, of course, vice versa.
- Change the size, color, margin, and other properties of your graph's titles (or any other text on the graph).
- Move your legend to another side of the graph, or even place it in the plot region.
- Change the aspect ratio of your graph.
- Stack the bars on a bar graph or turn them into percentages.
- Rotate or change the angle of axis labels.
- Add custom ticks and labels to the axes.
- Change the rule for the number and spacing of ticks and labels on an axis.
- Emphasize a point on the graph, whether marker, bar, spike, or other plot, by making it a custom color, size, or symbol.
- Change the text or properties of a marker label.

Because you can edit every property of every object on the graph, you can change almost anything about your graph. To learn more, see [G-1] **graph editor** or type `help graph editor`.

Notes

16 Saving and printing results by using logs

Using logs in Stata

When working on an analysis, it is worthwhile to behave like a bench scientist and keep a lab notebook of your actions so that your work can be easily replicated. Everyone has a feeling of complete omniscience while working intensely—this feeling is wonderful but fleeting. The next day, the exact small details needed for perfect duplication have become obscure. Stata has a lab notebook at hand: the *log* file.

A log file is simply a record of your Results window. It records all commands and all textual output as it happens. Thus it keeps your lab notebook for you as you work. Because it writes the file while it writes the Results window, it also protects you from disastrous failures, be they power failures or computer crashes. We recommend that you start a log file whenever you begin any serious work in Stata.

Logging output

All the output that appears in the Results window can be captured in a log file. Stata can save the file in one of two different formats. By default, Stata will save the file in its Stata Markup and Control Language (SMCL) format, which preserves all the formatting and links from the Results window. You can open these results in the Viewer, and they will behave as though they were in the Results window. If you would rather have plain-text files without any formatting, you can save the file as a plain log file. We recommend using the SMCL format, because SMCL files can be translated into a variety of formats readable by applications other than Stata with the **File > Log > Translate...** menu (see [R] **translate**).

To start a log file, click on the **Log** button, ![Log]. Choose **Begin...** from the drop-down menu to start a new log file. Choose **Append...** if you want to add more Stata output to an existing log file. Both choices open a standard file dialog that allows you to specify a directory and filename for your log. If you do not specify a file extension, the extension .smcl will be added to the filename.

Here is an example of a short session:

```
      name:  <unnamed>
       log:  /Users/mydir/Documents/base.smcl
  log type:  smcl
 opened on:  11 May 2011, 22:26:49
. sysuse auto
(1978 Automobile Data)
. by foreign, sort: summarize price mpg

-> foreign = Domestic
    Variable │      Obs        Mean    Std. Dev.       Min       Max

       price │       52    6072.423    3097.104        3291     15906
         mpg │       52    19.82692    4.743297          12        34

-> foreign = Foreign
    Variable │      Obs        Mean    Std. Dev.       Min       Max

       price │       22    6384.682    2621.915        3748     12990
         mpg │       22    24.77273    6.611187          14        41
. * be sure to include the above stats in report!
. * now for something completely different
. corr price mpg
(obs=74)
             │    price      mpg

       price │   1.0000
         mpg │  -0.4686   1.0000
. log close
      name:  <unnamed>
       log:  /Users/mydir/Documents/base.smcl
  log type:  smcl
  closed on:  11 May 2011, 22:34:46
```

There are a few items of interest.

- The header showing the log file's location, type, and starting timestamp is part of the log file. This feature helps when working with multiple log files.
- The two lines starting with asterisks (∗) are comments. Stata ignores the text following the asterisk, so you may type any comment you would like, with any special characters you would like. Commenting is a good way to document your thought process and to mark sections of the log for later use.
- In this example, the log file was closed using the log close command. Doing so is not strictly necessary because log files are automatically closed when you quit Stata.

Stata allows multiple log files to be open at once only if the log files are named. For details on this topic, see help log.

Working with logs

Log files are best viewed using Stata's Viewer. There are two ways to open a Viewer window:

- Click on the **Log** button, and select **View...**.
- Select **File > Log > View...**.

If there is a log file open (as shown by the status bar), it will be the default log file to view; otherwise, you need to either type the name of the log file into the dialog or click on the **Browse...** button to find the file with a standard file dialog.

Once you are in the Viewer window, everything behaves as expected: you can copy text and paste between the Viewer and anything else that uses text, such as word processors or text editors. You can even paste into the Command window or the Do-file Editor, but you should take care to copy only commands, not their output. It is okay to copy the prompt (".") at the start of the echoed command, because Stata is smart enough to ignore it in the Command window. When working with a word processor, what you paste will be unformatted text; it will look best if you use a fixed-width font, like Monaco, to display it.

Viewing your current log file is a good way to keep a reminder of something you have already done or a view of a previous result. The Viewer window takes a snapshot of your log file and hence will not scroll as you keep working in Stata. If you need to see more recent results in the Viewer, click on the **Refresh** button.

For more detailed information about logs, see [U] **15 Saving and printing output—log files** and [R] **log**. For more information about the Viewer, see [GSM] **3 Using the Viewer**.

Printing logs

To print a standard SMCL log file, you need to have the log file open in a Viewer window. Once the log file is in the Viewer, you can either click the **Print** button, right-click on the Viewer window, and select **Print...**, or you select **File > Print**.

- You can fill in none, any, or all of the items *Header*, *User*, and *Project*. You can check or uncheck options to *Print Line #'s*, *Print Header*, and *Print Logo*. These items are saved and will appear again in the print sheet (in this and in future Stata sessions).
- You can set the font size and color scheme the printer will use by clicking on the **Stata Headers and Footers** drop-down menu and choosing **Stata Fonts and Colors.** *Monochrome* is for black-and-white printing, *Color* is for default color printing, and *Custom 1* and *Custom 2* are for customized color printing.

You could also use the `translate` command to generate a PostScript or PDF file. See [R] **translate** for more information.

If your log file is a plain-text file (`.log` instead of `.smcl`), you can open it in a text editor, such as TextEdit, in the Do-file Editor in Stata, or in your favorite word processor. You can then edit the log file—add headings, comments, etc.—format it, and print it. If you bring the log file into a word processor, it will be displayed and printed with its default font. The log file will not be easily readable when printed in a proportionally spaced font (for example, Times Roman or Helvetica). It will look much better printed in a fixed-width font (for example, Monaco or Courier).

Rerunning commands as do-files

Stata also can log just the commands from a session without recording the output. This feature is a convenient way to make a do-file interactively. Such a file is called a *cmdlog* file by Stata. You can start a `cmdlog` file by typing

> `cmdlog using` *filename*

and you can close the `cmdlog` file by typing

```
cmdlog close
```
Here, for example, is what a `cmdlog` of the previous session would look like. It contains only commands and hence could be used as a do-file.

```
sysuse auto
by foreign, sort: summarize price mpg
* be sure to include the above stats in report!
* now for something completely different
corr price mpg
```

If you start working and then wish you had started a `cmdlog` file, you can save yourself heartache by saving the contents of the Review window. The Review window stores the last 5,000 commands you have typed. Simply right-click on the Review window and select **Save All...** from the menu. If you would like to move the commands directly to the Do-file Editor, select **Send to Do-file Editor**. You may find this method a more convenient way to create a text file containing only the commands that you typed during your session.

See [GSM] **13 Using the Do-file Editor—automating Stata**, [U] **16 Do-files**, and [U] **15 Saving and printing output—log files** for more information.

17 Setting font and window preferences

Changing and saving fonts and sizes and positions of your windows

You may find that you would like to change the fonts and display style of Stata's windows, depending on your monitor resolution and personal preferences. At the same time, there could be requirements for font usage, say, when you submit graphs to journals. Stata accommodates both of these by allowing sets of preferences for how windows are displayed.

We will first cover what can be changed in each window and then talk about what you can manage with your preferences.

Graph window

The preferences for the Graph window can be changed by right-clicking on the Graph window and choosing **Preferences...** from the contextual menu. The settings can then be set for graphs as they display in Stata. The settings that should be used when printing can be set under the **Printer** tab. The behavior of the Clipboard is controlled under the **Clipboard/PDF** tab.

The Graph preferences allow different schemes that control the look of graphs. These schemes provide a quick way to optimize graphs for printing or display on a screen. There are even schemes defined for *The Economist* and the *Stata Journal* so that you can get the details for these publications right without much fuss. Changing the scheme does not change the current graph—it applies the settings to future graphs.

All other windows

You can change the display font and font size for most types of windows in Stata.

When fonts and font sizes for a window can be changed, they can be changed by right-clicking on the window and selecting **Preferences...** from the contextual menu. Doing so will bring up the *Windows Preferences* dialog, from which you can pick the font and size of your choice. The font lists for each of the Results, Viewer, Data Editor, and Do-file Editor windows are restricted to fixed-width fonts only. This restriction ensures that output and numbers line up properly and are readable. The other windows can have any font that you would like without any adverse consequences.

Changing color schemes

The Results and Viewer windows have color schemes that control the way in which input, text, results, errors, links, and highlighted text display. Each has its color scheme set in the same fashion: you can right-click on the window and select or design your own color scheme. The default setting for both the Results window and the Viewer is the built-in *Standard* scheme which uses a white background and dark text. There are six other built-in schemes and three settings for custom schemes. The settings for the Viewer affect all Viewer windows at once.

Managing multiple sets of preferences

Stata's preferences are automatically saved when you quit Stata, and they are reloaded when Stata is launched. However, sometimes you may wish to rearrange Stata's windows and then revert to your preferred arrangement of windows. You can do this by saving your preferences to a named preference set and loading them later. Any changes you make to Stata's preferences after loading a preferences set do not affect the set; the set remains untouched unless you specifically overwrite it.

To manage preferences, open the **Stata > Preferences > Manage Preferences** menu, and do any of the following:

- Select **Manage Preferences...** to open the Manage Preferences window from which you can open, replace or delete an existing preference set.
- Select **Save Preferences...** to save the current window arrangement and preferences to disk. By default, these are saved in the Stata 12 Preferences folder in your Library. If you save your preferences to this default folder, they will appear in the **Stata > Preferences > Manage Preferences** menu the next time you view it.
- Select **Factory Settings** to restore all preferences to their original settings.
- Select **Factory Window Settings** to restore only the windowing preferences to their original settings.

Closing and opening windows

You can close all windows but the Results and Command windows. If you want to open a closed window, open the **Window** menu and select the desired window.

18 Learning more about Stata

Where to go from here

You now know plenty enough to use Stata. There is still much, much more to learn, because Stata is a rich environment for doing statistical analysis and data management. What should you do to learn more?

- Get an interesting dataset and play with Stata.
    a. Use the menus and dialog system to experiment with commands. Notice what commands show up in the Results window. You will find that Stata's simple and consistent command syntax will make the commands easy to read so that you will know what you have done, and easy to remember so that typing some commands will be faster than using menus.
    b. Play with graphs and the Graph Editor.
- If you venture into the Command window, you will find that many things will go faster. You will also find that it is possible to make mistakes where you cannot understand why Stata is balking.
    a. Try `help` *commandname* or **Help > Stata Command...** and entering the command name.
    b. Look at the command syntax and the examples in the help file, and compare them with what you typed. Compare them closely: small typographical errors make commands impossible for Stata to parse.
- Explore Stata by selecting **Help > Search...** and choosing *Search documentation and FAQs*. You will uncover many statistical routines that could be of great use. Explore beyond this by using the `findit` command.
- Look through the *Combined subject table of contents* in the *Quick Reference and Index*.
- Read and work your way through the *User's Guide*. It is designed to be read from cover to cover, and it contains most of the information you need to become an expert Stata user. It is well worth reading.
- Flip through the reference manuals, either paper or PDF, to read about statistical methods you like to use. The reference manuals are not meant to be read from cover to cover—they are meant to be referred to as you would an encyclopedia. You can find the datasets used in the examples in the manuals by selecting **File > Example Datasets...**, and then clicking on `Stata 12 manual datasets`. Doing so will enable you to work through the examples quickly.
- Stata has much information, including answers to frequently asked questions (FAQs), at http://www.stata.com/support/faqs/.
- There are many useful links to Stata resources at http://www.stata.com/links/resources.html. Be sure to look at these materials because many outstanding resources about Stata are listed here.
- Join Statalist, a listserver devoted to discussion of Stata and statistics.
- Read the official Stata blog Not Elsewhere Classified to read articles written by people at Stata about all things Stata. You can find the Stata blog at http://blog.stata.com.
- Visit Stata on Facebook at http://facebook.com/statacorp and follow Stata on Twitter at http://twitter.com/stata to keep up with Stata.
- If you prefer to sample the *User's Guide* and the references, there is some advice later in this chapter for you.

- Subscribe to the *Stata Journal*, which contains reviewed papers, regular columns, book reviews, and other material of interest to researchers applying statistics in a variety of disciplines. Visit http://www.stata-journal.com.
- Many supplementary books about Stata are available. Visit the Stata bookstore at http://www.stata.com/bookstore/.
- Take a Stata NetCourse®. NetCourse 101 is an excellent choice for learning about Stata. See http://www.stata.com/netcourse/ for course information and schedules.
- Attend a public training course taught by StataCorp at third-party sites. Visit http://www.stata.com/training/public.html for course information and schedules.

Suggested reading from the User's Guide and reference manuals

The *User's Guide* is designed to be read from cover to cover. The reference manuals are designed as references to be sampled when necessary.

Ideally, after reading this *Getting Started* manual, you should read the *User's Guide* from cover to cover, but you probably want to become at least somewhat proficient in Stata right away. Here is a suggested reading list of sections from the *User's Guide* and the reference manuals to help you on your way to becoming a Stata expert.

This list covers fundamental features and points you to some less obvious features that you might otherwise overlook.

Basic elements of Stata
  [U] **11 Language syntax**
  [U] **12 Data**
  [U] **13 Functions and expressions**

Data management
  [U] **6 Managing memory**
  [U] **21 Inputting and importing data**
  [D] **import** — Overview of importing data into Stata
  [D] **insheet** — Read text data created by a spreadsheet
  [D] **append** — Append datasets
  [D] **merge** — Merge datasets
  [D] **compress** — Compress data in memory

Graphics
  Graphics Reference Manual

Useful features that you might overlook
  [U] **28 Using the Internet to keep up to date**
  [U] **16 Do-files**
  [U] **19 Immediate commands**
  [U] **23 Working with strings**
  [U] **24 Working with dates and times**
  [U] **25 Working with categorical data and factor variables**
  [U] **13.5 Accessing coefficients and standard errors**
  [U] **13.6 Accessing results from Stata commands**
  [U] **26 Overview of Stata estimation commands**

[U] **20 Estimation and postestimation commands**

[R] **estimates** — Save and manipulate estimation results

Basic statistics

[R] **anova** — Analysis of variance and covariance

[R] **ci** — Confidence intervals for means, proportions, and counts

[R] **correlate** — Correlations (covariances) of variables or coefficients

[D] **egen** — Extensions to generate

[R] **regress** — Linear regression

[R] **predict** — Obtain predictions, residuals, etc., after estimation

[R] **regress postestimation** — Postestimation tools for regress

[R] **test** — Test linear hypotheses after estimation

[R] **summarize** — Summary statistics

[R] **table** — Tables of summary statistics

[R] **tabulate oneway** — One-way tables of frequencies

[R] **tabulate twoway** — Two-way tables of frequencies

[R] **ttest** — Mean-comparison tests

Matrices

[U] **14 Matrix expressions**

[U] **18.5 Scalars and matrices**

Mata Reference Manual

Programming

[U] **16 Do-files**

[U] **17 Ado-files**

[U] **18 Programming Stata**

[R] **ml** — Maximum likelihood estimation

Programming Reference Manual

Mata Reference Manual

System values

[R] **set** — Overview of system parameters

[P] **creturn** — Return c-class values

Internet resources

The Stata website (http://www.stata.com) is a good place to get more information about Stata. Half of the website is dedicated to user support. You will find answers to FAQs, ways to interact with other users, official Stata updates, and other useful information. You can also subscribe to Statalist, a listserver devoted to discussion of Stata and statistics.

You will also find information on Stata NetCourses®, which are interactive courses offered over the Internet that vary in length from a few weeks to eight weeks. Stata also offers in-person training sessions. Visit the Stata website for more information.

At the website is the Stata Bookstore, which contains books that we feel may be of interest to Stata users. Each book has a brief description written by a member of our technical staff explaining why we think this book may be of interest.

We suggest that you take a quick look at the Stata website now. You can register your copy of Stata online and request a free subscription to the *Stata News*.

Visit http://www.stata-press.com for information on books, manuals, and journals published by Stata Press. The datasets used in examples in the Stata manuals are available from the Stata Press website.

Also visit http://www.stata-journal.com to read about the *Stata Journal*, a quarterly publication containing articles about statistics, data analysis, teaching methods, and effective use of Stata's language.

Visit Stata's official blog at http://blog.stata.com for news and advice related to the use of Stata. The articles appearing in the blog are individually signed and are written by the same people who develop, support, and sell Stata. The Stata Blog also has links to other blogs about Stata, written by Stata users around the world.

Follow Stata on Twitter at http://twitter.com/stata and on Facebook at http://facebook.com/statacorp. These are both good ways to stay up-to-the-minute with the latest Stata information.

See [GSM] **19 Updating and extending Stata—Internet functionality** for details on accessing official Stata updates and free additions to Stata on the Stata website.

19 Updating and extending Stata—Internet functionality

Internet functionality in Stata

Stata works well together with the Internet. Stata can use datasets and view remote help files as though they were on your computer. Stata also can keep itself up to date (with your permission, of course). Finally, you can install *user-written commands*, which are commands that extend Stata's functionality. These are commands that have been presented in the *Stata Journal* (SJ), the *Stata Technical Bulletin* (STB), or have simply been written and shared by the greater Stata community.

This chapter will show you how you can expand Stata's horizons.

Using files from the Internet

Stata understands URLs as though they were local file locations. If you know of a file on the web that you would like to use, be it a dataset, a graph, or a do-file, you can easily open it in Stata. Here is a small example.

There are many datasets at http://www.stata-press.com/data/. Suppose that you would like to use the census12 dataset used in [U] **11 Language syntax**, and you know that its location is at http://www.stata-press.com/data/r12/census12.dta. Because you know that the command for opening a dataset is use, you could type the following:

```
. use http://www.stata-press.com/data/r12/census12.dta
(1980 Census data by state)

. describe

Contains data from http://www.stata-press.com/data/r12/census12.dta
  obs:            50                          1980 Census data by state
 vars:             7                          6 Apr 2011 15:43
 size:         1,950

              storage   display    value
variable name   type    format     label      variable label

state          str14    %14s                  State
state2         str2     %-2s                  Two-letter state abbreviation
region         str7     %9s                   Census region
pop            long     %10.0g                Population
median_age     float    %9.2f                 Median age
marriage_rate  float    %9.0g                 Marriage rate
divorce_rate   float    %9.0g                 Divorce rate

Sorted by:
```

This functionality is everywhere in Stata. Any command that reads a file with a *filename* in its syntax can use a web address as easily as a file that is stored on your computer.

Official Stata updates

By official Stata, we mean the pieces of Stata that are provided and supported by StataCorp. The other and equally important pieces are the user-written additions published in the SJ, distributed over Statalist, or distributed in other ways.

Stata can fetch both official updates and user-written programs from the Internet. Let's start with the official updates. StataCorp often releases updates to official Stata. These updates add new features and, sometimes, fix bugs.

By default, Stata has automatic update checking turned on and set to check for updates every seven days. To change or check your settings, select **Stata > Preferences > General Preferences...**.

We recommend using automatic update checking because it is a simple, unobtrusive way to be sure that your copy of Stata is always up to date. If you keep this default, you will be prompted with a dialog when you start Stata if you have not recently checked for updates.

To manually check whether there are any official Stata updates, either click on **Help > Check for Updates** or type `update query` in the Command window. Regardless of which choice you make, Stata goes to check for official updates. After it checks, it will show you your update status. If your copy of Stata is already up-to-date you will be told. If your copy of Stata needs updating, you will be told, and a link, `Install available updates`, will show up in your Results window. You can click on this link or type `update all` and press *Return*. In either case, Stata will download what is needed to bring your copy of Stata up to date. Stata will need to restart after being updated, so it gives you get a chance to postpone the update in case there was something (such as saving the command history) you wanted to do in the current session.

Troubleshooting note: If you do not have write permission for `/Applications/Stata`, you cannot install official updates in this way. You may still download the official updates, but you will need to use the command-line version of `update`; see [U] **28 Using the Internet to keep up to date** for instructions.

Automatic update checking

Stata can periodically check for updates for you. By default, Stata will check once every seven days for updates from the StataCorp website. The seven-day interval is from the last time an `update query` was performed regardless of whether it was by Stata or by you. You can change the interval between checks.

Before Stata connects to the Internet to check for an update, it will ask you if you would like to check now, check the next time Stata is launched, or check after the next interval. You can disable the prompt and allow Stata to check without asking.

If an update is available, Stata will notify you. From there, you should follow the recommendations for updating Stata.

You can change the settings for automatic update checking by selecting **Stata > Preferences > General Preferences...** and choosing **Internet**.

Finding user-written programs by keyword

Stata has a built-in utility created specifically to search the Internet for user-written Stata programs. You can access it by selecting **Help > Search...**, choosing *Search net resources*, and entering a keyword in the field. Choosing **Help > SJ and User-written Programs** yields more specific choices for searching. The utility searches all user-written programs on the Internet, including the entire collection of *Stata Journal* and STB programs. The results are displayed in the Viewer, and you can click to go to any of the matches found.

For the syntax on how to use the equivalent search *keywords*, net command, see [R] **search**.

Downloading user-written programs

Downloading user-written programs is easy. Start by selecting **Help > SJ and User-written Programs**:

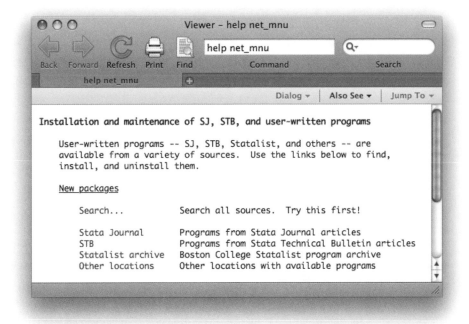

As the Viewer says, try Search... first.

Suppose that you were interested in finding more information or some user-written programs involving cubic splines. You select **Help > Search...**, select *Search all*, type cubic spline in the search box, and click on the **OK** button.

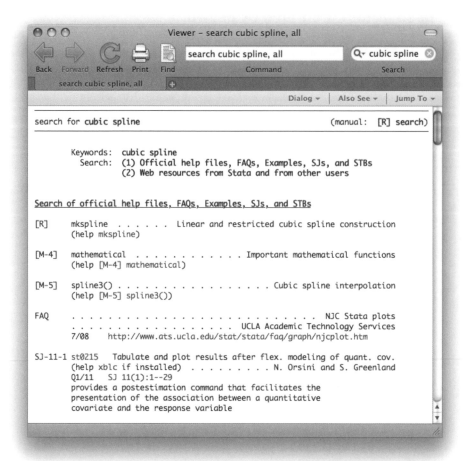

The first entry points to the built-in Stata command mkspline. You investigate this command and find it interesting. You see that the next two entries point to some built-in routines in Mata. You follow these links because Mata is not only intriguing, but also fast. You see that the next link points to an FAQ on UCLA's website. Finally, you decide to check the fifth link. It points to an article in the *Stata Journal*, volume 11, number 1 (first quarter, 2011). You should click on the st0215 link, because it will go to the programs associated with this article.

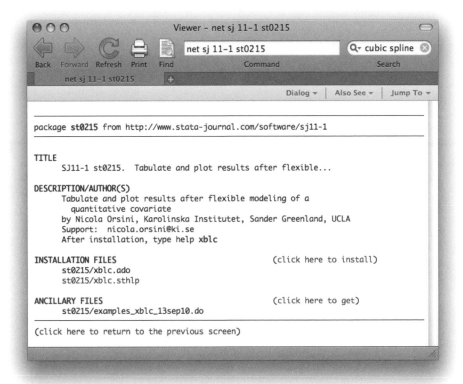

You can click on the help file (st0215/xblc.sthlp) to see if the one command here looks interesting. If you decide that you would like to try the command, you can click on the link click here to install. If you decide that you would like to use some of the ancillary files—files that typically help explain the workings of the command, you could download those, too. You do not need to worry—doing so will not interfere in any way with your copy of Stata. We will show you how to safely uninstall these programs shortly. Click on the install link to install the package. That is all there is to installing a user-written package.

Now suppose that you decide that you would like to uninstall the package. Doing so is simple enough: select **Help > SJ and User-written programs**, and click on the List link. You should see the following:

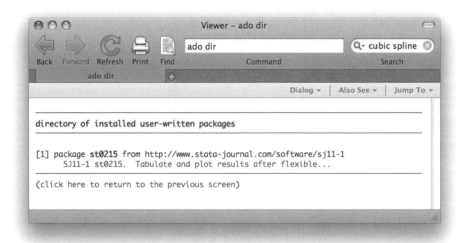

If you click on the one-line description of the program, you will see the full description of what has been installed. Here is a short version of what you would see:

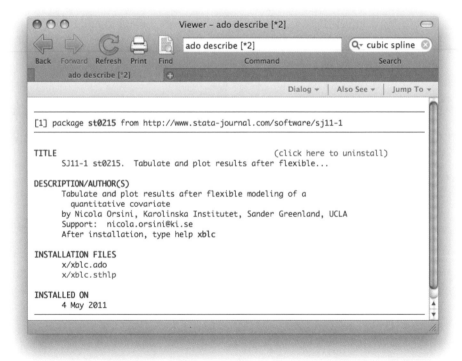

You can uninstall materials by clicking on click here to uninstall when you are looking at the package description. Try it.

For information on downloading user-written programs by using the net command, see [R] net.

A Troubleshooting Stata

Contents

A.1 If Stata does not start

You tried to start Stata and it refused; Stata or your operating system presented a message explaining that something is wrong. Here are the possibilities:

Cannot find license file

This message means just what it says; nothing is too seriously wrong. Stata simply could not find the license file it was looking for. The most common reason for this is that you did not complete the installation process.

Did you enter the codes printed on your paper license to unlock Stata? If not, go back and complete the initialization procedure.

Error opening or reading the file

Something is distinctly wrong for purely technical reasons. Stata found the file that it was looking for, but there was an I/O error.

About the only way this situation could arise would be a hard-disk error. Stata technical support will be able to help you diagnose the problem; see [U] **3.9 Technical support**.

License not applicable

Stata has determined that you have a valid Stata license, but the license does not apply to the version of Stata that you are trying to run.

The most common reason for this message is that you have a license for Stata/IC but you are trying to run Stata/SE or Stata/MP, or you have a license for Small Stata but you are trying to run Stata/IC, Stata/SE, or Stata/MP. If any of these is the case, reinstall Stata, making sure to choose the appropriate flavor.

You can't open the application StataMP because it is not supported on this architecture

You are trying to run Stata/MP on a 32-bit computer. Stata/MP for Mac runs only on 64-bit computers.

Other messages

The other messages indicate that Stata thinks you are attempting to do something that you are not licensed to do. Most commonly, you are attempting to run Stata over a network when you do not have a network license, but there are many other alternatives. There are two possibilities: either you really are attempting to do something that you are not licensed to do, or Stata is wrong. In either case, you are going to have to contact us. Your license can be upgraded, or, if Stata is wrong, we can provide codes to make Stata stop thinking that you are violating the license; see [U] **3.9 Technical support**. You also could be trying to run Stata on Mac OS X 10.4 or earlier. The minimum requirement for Stata is Mac OS X 10.5.6.

A.2 Troubleshooting tips

Stata has a minimum requirement of a Mac computer (PowerPC or Intel) running Mac OS X 10.5 or better.

If you experience a crash while launching Stata, open the `Preferences` folder in your home directory's `Library` folder. Trash the file `com.stata.stata12.plist`, and restart Stata. It is an uncommon event, but the preference file can become corrupted by previous crashes caused by power outages or by not shutting down.

Also see the frequently asked questions (FAQs) for Mac on the Stata website at http://www.stata.com/support/faqs/mac. The solution to your problem may be there.

If you want us to help, we will need as much information as possible. Restart Stata and reproduce the problem, writing down everything you do before the problem occurs.

Also tell us about your computer. What model Mac is it? What version of the system software are you using? Click on **Apple Menu > About This Mac**, and then click on the **More Info...** button to open the System Profiler application. You can generate a profile of your system that you can email to us. Finally, we need your Stata serial number and the revision date of your version of Stata. Include them if you email, and know them if you call. You can obtain them by typing the `about` command in Stata's Command window. `about` lets you know everything about your copy of Stata, including the version and the date it was produced.

B Managing memory

Contents

B.1 Memory size considerations

Starting in Stata 12, memory management in Stata is automatic—there is no longer any requirement to manually request memory for Stata when you know you will be using a large dataset. For details on efficiency tweaks needed by a very few Stata users, look at [D] **memory**.

Notes

C   Advanced Stata usage

Contents

C.1 Executing commands every time Stata is started

Stata looks for the file `profile.do` when it is invoked and, if it finds it, executes the commands in it. Stata looks for `profile.do` first in the directory where Stata is installed, then in the current directory, then along your path, and finally along the ado-path (see [P] **sysdir**); we recommend that you put `profile.do` in /Users/*yourhome*/Library/Application Support/Stata.

Say that every time you start Stata, you want `matsize` set to 800 (see [R] **matsize**). In /Users/*yourhome*/Library/Application Support/Stata, create the file `profile.do` containing

```
set matsize 800
```

When you invoke Stata, the usual opening appears, but with the following additional command, which will be executed:

```
running ~/Library/Application Support/Stata/profile.do ...
```

You could also type `set matsize 800, permanently` in Stata. The `permanently` option tells Stata to remember the setting for future sessions and eliminates the need to put the `set matsize` command in `profile.do`.

`profile.do` is treated just as any other do-file once it is executed; results are just the same as if you had started Stata and then typed `run profile.do`. The only special thing about `profile.do` is that Stata looks for it and runs it automatically.

System administrators might also find `sysprofile.do` useful. This file is handled in the same way as `profile.do`, except that Stata first looks for `sysprofile.do`. If that file is found, Stata will execute any commands it contains. After that, Stata will look for `profile.do` and, if that file is found, execute the commands in it.

One example of how `sysprofile.do` might be useful would be when system administrators want to change the path to one of Stata's system directories. Here `sysprofile.do` could be created to contain the command

```
sysdir set SITE "/Library/Application Support/Stata"
```

See [U] **16 Do-files** for an explanation of do-files. They are nothing more than ASCII text files containing sequences of commands for Stata to execute.

C.2 Other ways to launch Stata

You can start Stata by double-clicking on a Stata `.dta` dataset, a Stata `.do` do-file, or a Stata `.gph` graph file. In all cases, your current working directory will become the folder containing the file you have double-clicked.

Stata will behave as you would expect in each case. If you double-click on a dataset, Stata will open the dataset after Stata starts. If you double-click on a do-file, the do-file will be opened in the Do-file Editor. If you double-click on a graph, the graph will be opened by Stata.

If you would rather have Stata execute the commands in a do-file when it is double-clicked, select **Stata > Preferences**, click on the **Do-file Editor** toolbar button, click on the **Advanced** tab, and uncheck the *Edit do-files opened from the Finder in Do-file Editor* checkbox.

C.3 Stata batch mode

To run Stata in batch mode, you need to start it in the Terminal. The syntax of the command to start Stata from in the Terminal is

$$\text{StataSE } \begin{bmatrix} \textit{-option} \begin{bmatrix} \textit{-option} \begin{bmatrix} \ldots \end{bmatrix} \end{bmatrix} \end{bmatrix} \begin{bmatrix} \textit{stata\_command} \end{bmatrix}$$

where the options are

Option	Result
-b	set background (batch) mode and log in ASCII text
-e	set background (batch) mode and log in ASCII text without prompting when Stata command has completed
-q	suppress logo and initialization messages
-s	set background (batch) mode and log in SMCL

The -q option starts Stata, but suppresses all the initialization messages, including the Stata logo.

For you to run Stata from the Terminal, you need to be sure that the shell can find Stata. To do this, you must add the path to the Stata executable in Stata's application bundle to your shell's path. Once that is done, you can invoke Stata from any directory from a shell.

For example, if Stata is installed in /Applications/Stata, then the path to the executable for Stata/SE is /Applications/Stata/StataSE.app/Contents/MacOS. Type StataSE to start Stata/SE.

For Stata/MP, it is /Applications/Stata/StataMP.app/Contents/MacOS. Type StataMP to start Stata/MP.

For Stata/IC, it is /Applications/Stata/Stata.app/Contents/MacOS. Type Stata to start Stata/IC.

For Small Stata, it is /Applications/Stata/smStata.app/Contents/MacOS. Type smStata to start Small Stata.

Suppose you had a do-file named bigjob.do. If you want to use Stata in batch mode, typing

```
% StataSE -b do bigjob
```

tells Stata to execute the commands in bigjob.do, suppress all screen output, and route the output to bigjob.log in the same directory. Stata will display a dialog when the commands have finished executing.

Typing

```
% StataSE -e do bigjob
```

tells Stata to execute the commands in bigjob.do, suppress all screen output, and route the output to bigjob.log in the same directory. Stata will simply exit without displaying a dialog when the commands have finished executing.

```
% StataSE -s do bigjob
```

tells Stata to execute the commands in `bigjob.do`, suppress all screen output, and route the output to `bigjob.smcl` in the same directory.

You can also run the above examples in the background by typing

```
% StataSE -b do bigjob &
% StataSE -e do bigjob &
% StataSE -s do bigjob &
```

Note: Stata runs `profile.do` before doing `bigjob.do`, just as it would if you were working interactively.

Notes

D More on Stata for Mac

D.1 Using Stata datasets and graphs created on other platforms

Stata will open any Stata `.dta` dataset or `.gph` graph file, regardless of the platform on which it was created, even if it was a Windows or Unix system. Also, Stata for Windows and Stata for Unix users can use any files that you create. Remember that `.dta` and `.gph` files are binary files, not text (ASCII) files, so they need no translation; simply copy them over to your hard disk as is.

Files created on other platforms can be directly opened from the Stata command line; for example, you can load a dataset by typing `use` *filename* or by double-clicking on the file in the Finder.

D.2 Exporting a Stata graph to another document

Suppose that you wish to export a Stata graph to a document created by your favorite word processor or presentation application. You have two main choices for exporting graphs: you may copy and paste the graph by using the Clipboard, or you may save the graph in one of several formats and import the graph into the application.

D.2.1 Exporting the graph by using the Clipboard

The easiest way to export a Stata graph into another application is to use drag and drop. You can select the graph in the Graph window and drag it to the location you would like it in another open application.

You can also export graphs by copying and pasting.

Either create your graph or redisplay an existing graph. To copy it to the Clipboard, right-click on the Graph window, and select **Copy**. Stata will copy the graph in PDF format. This is the native graphics format for Mac OS X. Programs made to edit vector graphics can be used to edit the elements of the graph.

After you have copied the graph to the Clipboard, switch to the application into which you wish to import the graph and paste it. In most applications, this is accomplished by selecting **Edit > Paste**. Some applications require you to first create a bounding box for the image to be placed in. If pasting does not immediately work, see your application's documentation on pasting images from the Clipboard into documents.

D.2.2 Exporting the graph to a file

Stata can export graphs to several different file formats. If you right-click on a graph, select **Save As...**, and then click on *Format*, you will see that Stata can save in the following file types: PDF, JPEG, Portable Network Graphics (PNG), TIFF, PostScript, Enhanced PostScript (EPS), and EPS with TIFF preview. PDF, PostScript, and EPS formats are vector formats, whereas JPEG, PNG and TIFF are bitmap formats. EPS files are recommended when you want to export your graph to an application on another platform, such as TEX or LATEX on Unix (which is how all the graphs in Stata's manuals were created), or for best output on a PostScript printer. If you wish to include a thumbnail of the graph, choose **EPS with TIFF preview**. Choosing the preview option does not affect how the graph is printed. PNG is well suited for placing graphs on a webpage. See the *Graphics Reference Manual* for more information.

D.3 Stata and the Notification Manager

You may work in another application while Stata is processing a time-intensive command or do-file. Stata will notify you when it has finished by playing a sound and bouncing its application icon on the Dock. The default notification behavior is for Stata to bounce its application icon on the Dock once. The notification feature works only if Stata is put into the background while it is executing a command or do-file. Stata will not notify you if it is in the foreground or requires further interaction from you (such as a —more— condition). You can control how notifications work in Stata's preferences.

D.4 Setting the default end-of-line delimiter

The end-of-line delimiter for text files on Mac OS X and Unix systems is a line-feed character. You can change the end-of-line delimiter Stata for Mac uses to output to text files to the old Mac OS end-of-line delimiter by typing

```
set eolchar mac
```

You can change it back to a Mac OS X end-of-line delimiter by typing

```
set eolchar unix
```

You can make either change permanent by using the `permanently` option:

```
set eolchar unix, permanently
```

D.5 Stata(console) for Mac OS X

D.5.1 Requirements

Stata(console) is included with both Stata/SE and Stata/MP for the Mac. It runs in a Terminal window without a graphical user interface (GUI)—there are no Data Editor, Viewer, or Graph windows. Graphs and datasets can be saved, as usual, they simply cannot be viewed interactively. Stata(console) is meant for running Stata remotely and for running batch jobs in the background.

You can also run background jobs by using your standard Stata installation. The command line options are given in [GSM] **C.4 Stata batch mode**.

You must already have Stata/SE or Stata/MP for the Mac installed and your license initialized before installing Stata(console). If you have a single-user license and wish to have more than one login ID use Stata at a time, please contact our sales department to purchase an upgrade to a multiuser license. You should have some experience in working from a shell in Unix before attempting to set up Stata(console). You must also have administrator access to your computer to complete the setup. With administrator access, there is always the potential damaging your computer.

In the instructions that follow, use `stata-mp` in place of `stata-se` if you are installing the console version of Stata/MP.

D.5.2   Installing Stata(console) using the Terminal utility

If you have installed Stata in its default location `/Applications/Stata`, Stata(console) can be installed by selecting **Stata > Install Terminal Utility...**. You can then start the console version by typing `stata-se` or `stata-mp` in a Terminal window.

If you rename Stata's folder or move it out of the Applications folder after installing the utility, the symbolic link to the console application will be broken. In this case, just reinstall the utility by selecting **Stata > Install Terminal Utility...** again.

D.5.3   Installing Stata(console) by hand

If you would rather install Stata(console) by hand instead of using the built-in utility, read this section. If you are not comfortable with Unix or are wary of damaging your computer, skip the rest of this section. If you do not know when to use

```
setenv PATH /Applications/Stata:$PATH
```

and when to use

```
PATH=/Applications/Stata:$PATH
```

you should not attempt the installation.

1. Log in to your Mac with an account that has administrator privileges.
2. Install Stata/SE or Stata/MP in */Applications/Stata* and initialize your license if Stata is not already installed.

You have completed the installation process. We now wish to verify that everything is installed properly as a regular user. To use Stata as a regular user, your path must include */Applications/Stata/StataSE.app/Contents/MacOS*. For a quick test, we are going to be crude.

```
mymac: > export PATH=$PATH:/Applications/Stata/StataSE.app/Contents/MacOS
mymac: > stata-se
```

Stata should come up. Once you have verified that everything is working, make sure that your users modify their shell startup script to include */Applications/Stata/StataSE.app/Contents/MacOS* in their path. If they do not have a `.bash_profile` file already, they should make one and include the following line:

```
export PATH=$PATH:/Applications/Stata/StataSE.app/Contents/MacOS
```

D.5.4   Updating Stata(console) in the future

Stata(console) is automatically updated when the regular Stata(GUI) application is updated. All updating should be performed from within Stata(GUI).

Subject index

This is the subject index for the *Getting Started with Stata for Mac* manual. Readers may also want to consult the combined subject index (and the combined author index) in the *Quick Reference and Index*.